T0135656

Gerd Kunert

Advances in
a posteriori error estimation on
anisotropic finite element discretizations

Von der Fakultät für Mathematik der Technischen Universität Chemnitz angenommene
Habilitationsschrift zur Erlangung des akademischen Grades doctor rerum naturalium
habilitatus (Dr. rer. nat. habil.)

Bibliografische Information Der Deutschen Bibliothek

Die Deutsche Bibliothek verzeichnet diese Publikation in der Deutschen Nationalbibliografie; detaillierte bibliografische Daten sind im Internet über http://dnb.ddb.de abrufbar.

ISBN 3-8325-0450-8

Logos Verlag Berlin
Comeniushof, Gubener Str. 47,
10243 Berlin
Tel.: +49 030 42 85 10 90
Fax: +49 030 42 85 10 92
INTERNET: http://www.logos-verlag.de

Preface

Certain classes of partial differential equations generically give rise to solutions with strong directional features, e.g. with boundary layers. Such solutions are called *anisotropic*. Their discretization by means of the finite element method (for example) can favourably employ so-called *anisotropic meshes*. These meshes are characterized by stretched, anisotropic finite elements with a (very) large stretching ratio.

The widespread use of computer simulation leads to an increasing demand for semi- or fully automatic solution procedures. Within such self-adaptive algorithms, *a posteriori* error estimators form an indispensable ingredient for quality control. They are well understood for standard, isotropic discretizations.

The knowledge about *a posteriori* error estimation on anisotropic meshes is much less mature. During the last decade the foundation and basic principles have been proposed, discussed and established, mostly for the Poisson problem. This monograph summarises some of the recent advances in anisotropic error estimation for more challenging problems. Emphasis is given to the contributions of the author.

In Chapter 3 the investigation starts with *singularly perturbed reaction diffusion problems* which frequently lead to solutions with boundary layers. This problem class often arises when simplifying more complex models. Chapter 4 treats *singularly perturbed convection diffusion problems*, i.e. the convection is dominating. The solution structure is more intricate, and often features boundary layer and/or interior layer solutions. Chapter 5 is devoted to the *Stokes equations*. Flow problems generically give rise to anisotropic solutions (e.g. with edge singularities or containing layers). The Stokes equations often serve as a simplified or linearised model. In all three chapters, the main results consist in error estimators and corresponding error bounds that are robust with respect to the mesh anisotropy, as far as possible.

Finally Chapter 6 addresses the robustness of *a posteriori* error estimation with respect to the mesh anisotropy. In particular the relation between anisotropic mesh construction and error estimation is investigated.

This thesis presents the philosophy of anisotropic error estimation as well as the main *results* and the definitions required. Proofs and technical details are omitted; instead the key ideas are explained. The compact style of presentation aims at practitioners in particular by providing easily accessible error estimators and error bounds. Further insight is readily possible through the references.

I would like to thank my collaborators Serge Nicaise and Emmanuel Creusé; our joint work has led to the results for the Stokes problem. I am indebted to my colleagues for stimulating discussions and comments. Thomas Apel, Bernd Heinrich and Arnd Meyer deserve special thanks. Finally this work has been possible only with the encouragement from my family.

Chemnitz, March 2003

Contents

Copyright notice

Most of the material of this book has been published already as journal articles (the presentation here is always adapted and modified). Therefore we would like to thank the following publishers for the permission to reproduce the material. The copyright is as indicated below.

Chapter 1

Introduction

1.1 Motivation of anisotropic finite elements

Today computer simulation is a wide-spread and well accepted tool to investigate physical phenomena and engineering problems. The fields of application are as diverse as automobile construction, weather and climate prediction, computer tomography in medicine, pollution forecasts, financial mathematics, molecule design and analysis (in particular in life science industries and biochemistry), or aircraft design, to name but a few. The modelling often leads to partial differential equations (PDEs). Among the numerical solution schemes, the *Finite Element Method (FEM)* is one of the most successful and powerful approaches. It has been proposed and founded in the middle of the last century, and has gained enormous popularity afterwards. Today it is well established and analysed in theory and practice, and many specific variants are known and used. We assume that the reader is familiar with the basic concept of the FEM, cf. also the standard textbook [Cia78]. The *mesh* (or *triangulation*) is denoted by $\mathcal{T}$. It consists of *elements* T which can, for example, be triangles or quadrilaterals (for two dimensional problems), or tetrahedra, pentahedra, or hexahedra (for three dimensional problems).

In this work we solely consider the finite element method. As the title suggests, two aspects will be of particular interest, namely

- anisotropic discretizations, and
- error estimation.

Special attention is paid to the combination of both items. Nevertheless we start with a separate description.

Anisotropic finite element discretizations

Certain problem classes frequently give rise to solutions with strong directional features. In a mathematical (or physical) diction these problems include

- the Poisson problem in 3D domains with concave (i.e. re-entrant) edges,
- reaction diffusion problems with small diffusion,
- convection diffusion problems with dominating convection,
- plate problems (with a sufficiently accurate model, e.g. Reissner Mindlin),
- Stokes and Navier-Stokes problems,
- microconductor device simulation.

1

An illustrative example of a solution with directional behaviour is given by the shock waves that are induced by the (supersonic) flow around an airplane. More general, a directional solution shows much variation in one spatial direction but little variation in an other direction. In this sense the solution exhibits an (almost) lower dimensional behaviour.

Such directional solutions are termed *anisotropic*. For a better visualisation Figure 1.1.1 displays a very simple anisotropic function (in a 2D domain) whose anisotropy is given by two (exponential) boundary layers.

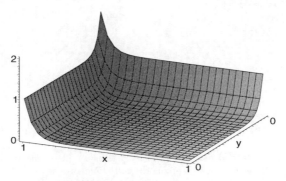

Figure 1.1.1: Example of an anisotropic function with two (exponential) boundary layers.

Let us now turn to an appropriate finite element *discretization* of a problem with an anisotropic solution. The example of Figure 1.1.1 will be used to illustrate the ideas.

Isotropic and uniform mesh: The earliest FEM implementations employed *uniform* meshes. There all elements are of similar size (and possibly even identical), cf. the left part of Figure 1.1.2. The notion *isotropic* means that the ratio of the diameters of the circumscribed sphere and the inscribed sphere is bounded from above for each element. For triangles this is equivalent to the demand that the minimum angle of the triangle is bounded away from 0.

Such a uniform discretization is comparatively easy to implement and analyse. Unfortunately this variant is little effective or even useless when the solution shows strongly different behaviour in different parts of the computational domain. This is frequently the case for anisotropic solutions, see again Figure 1.1.1.

Isotropic and adapted mesh: When a solution behaves differently in certain parts of the domain, engineers and researchers soon tried to accommodate the mesh to this behaviour. In regions where the solution shows much variation a fine discretization is appropriate. In terms of the FEM, the elements will be small there. Conversely, smooth parts of the solution can be approximated by a coarse discretization (corresponding to large elements in the FEM). In this way one can obtain an accuracy comparable to that of a uniform mesh but with (much) less elements (and thus computational cost). The right part of Figure 1.1.2 provides a lively impression of this approach.

In practice one wants to construct the appropriately adapted meshes more or less automatically. This can be achieved by adaptive algorithms that produce a sequence of adapted meshes. The key idea is to *refine* the mesh in those parts where the numerical error is large. There is a vast literature on this topic; here we refer to [Ver96, AO00] and the citations therein.

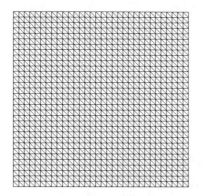 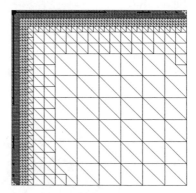

Figure 1.1.2: Examples of isotropic meshes.
Left: Uniform mesh   Right: Adapted mesh

Anisotropic and adapted mesh: In the special situation of an anisotropic solution it is natural to reflect the anisotropy by a proper discretization. In contrast to an isotropic mesh, one may vary not only the *size* of the elements but also their *shape*. The directional features of the solution are reflected by *stretched elements*. Such elements are called *anisotropic*. The *stretching ratio* or *aspect ratio* is defined as the ratio of the diameters of the smallest circumscribed sphere and the largest inscribed sphere. In contrast to the conventional isotropic theory, this stretching ratio in no longer bounded but can become arbitrarily large instead.

Beside the stretching ratio, the *stretching direction* is equally important. Intuitively the elements should be stretched in that spatial direction where the anisotropic solution shows little variation. Figure 1.1.3 gives two examples of appropriate anisotropic meshes for discretizing the anisotropic solution of Figure 1.1.1.

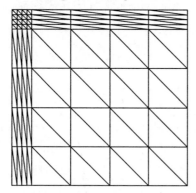 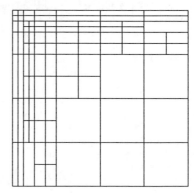

Figure 1.1.3: Examples of anisotropic meshes.
Left: Triangular mesh (structured; constructed *a priori*)
Right: Rectangular mesh (adaptively constructed)

In comparison with isotropic meshes one can now achieve a comparable accuracy with less elements and less computational effort. Anisotropic discretizations can be particularly advantageous (or even necessary) for 3D problems that require many elements (whose number may easily exceed a million).

Anisotropic discretizations have gained increasing popularity within the last one or two decades. Exemplarily we refer to [PVMZ87, KR90, AD92, ZR93, Rac93, RGK93, ZW94, BK94, Noc95, FLR96, Sie96, VH96, RST96, MOS96, AL96, BD97, HL98, Roo98, AL98, Ape99, Kun99, RL99, DGP99, SR99, Fle99, KV00, FPZ01, ANS01a, ANS01b, Kra01, Gro02, Lin03, Pic03]. This incomplete list of investigations documents the intensified interest in anisotropic elements but also the difficulties associated with this approach. The papers range from applications of anisotropic discretizations to purely analytical approaches.

Error estimation and adaptive algorithms

The previous paragraphs have motivated the advantageous use of anisotropic finite elements. One of the major questions is how to obtain and construct anisotropic FEM discretizations adaptively. In order to answer this question and to work out the technicalities that are associated with anisotropic elements, we start with a recapitulation of *isotropic* discretizations. A standard isotropic adaptive algorithm consists basically of the following steps.

Isotropic **adaptive algorithm:**

0. Start with an initial mesh $\mathcal{T}_0$.
1. Solve the corresponding discrete system.
2. Compute the local *a posteriori* error estimator for each element T of the mesh.
3. When the estimated global error is small enough then stop.
   Otherwise obtain information for a new, better mesh, namely the element size (as a function over Ω).
4. Based on this information, construct a new mesh or perform a mesh refinement, and re-iterate with step 1.

In contrast, an adaptive algorithm that aims at *anisotropic* discretizations contains the following basic ingredients.

Anisotropic **adaptive algorithm:**

0. Start with an initial mesh $\mathcal{T}_0$.
1. Solve the corresponding discrete system.
2. Compute the local *a posteriori* error estimator for each element T of the mesh.
3. When the estimated global error is small enough then stop.
   Otherwise obtain information for a new, better mesh. This includes:
   - Detect regions of anisotropic behaviour of the solution.
   - Determine a (quasi) optimal stretching direction and stretching ratio of the finite elements in those regions.
   - Determine the element size.
4. Based on this information, construct a new mesh or perform a mesh refinement, and re-iterate with step 1.

Let us now examine the differences between the isotropic and the anisotropic version. Obviously step 4 (mesh refinement) is different, since now anisotropic elements are to be generated. The information extraction of step 3 also is more extensive, since now anisotropic, additional information is required. Chapter 6 deals with this topic; for the moment we don't go into more details.

The solution of the discrete system of equations (step 1) needs reinvestigation too. Every class of problems requires its own solver and/or preconditioner that is tailored to perform well on anisotropic discretizations. Partially these tasks have already been addressed. Here we do not consider this topic, as it is beyond our scope.

Our main focus, however, is on the *a posteriori* error estimation of step 2. Many standard estimators fail when they are applied on anisotropic elements. To be more precise, we aim at bounds for the error from above and below (which are closely related to *reliability* and *efficiency* of the error estimator). In general these bounds depend on the maximum stretching ratio of all elements of the mesh. This dependence is harmless for *isotropic* meshes; there the stretching ratio is always $\mathcal{O}(1)$, and upper and lower error bounds differ (at most) by some constant. For *anisotropic* meshes the situation is completely different. The maximum aspect ratio may be very large (e.g. easily exceeding 1000), and upper and lower error bounds can differ by the same factor. Hence the error bounds would be absolutely useless in practice.

Our main concern therefore is to define and analyse *a posteriori* error estimators that provide useful and tight error bounds on anisotropic discretizations. This task is important not only for adaptive algorithms but also for (stand alone) control of the quality of an approximate solution. Chapters 3–5 are devoted to this error estimation.

Before the content of our work is described in detail, we have to distinguish between two classes of error estimates, namely *a priori* and *a posteriori* estimates.

- *A priori* estimates provide bounds on the discretization error in terms of the given data of the PDE on a family of *a priori* defined meshes. The bounds may contain the exact solution u of the PDE but *not* the approximate solution u_h.

  A priori error bounds are valuable for the development and theoretical justification of solution methods. For example, optimally graded meshes can be devised for anisotropic solutions with edge singularities, cf. [Ape99]. *A priori* error bounds are in general unsuitable for adaptive algorithms or quality control.

- *A posteriori* estimates use the data of the PDE and the approximate solution u_h but not the exact solution u, cf. the textbooks [Ver96, AO00] (for isotropic discretizations). These estimates can be used for quality control or adaptive algorithms. Here we solely investigate this type of error estimation.

The aims of both kinds of error estimation are similar. Nevertheless their respective analysis employs quite different techniques and resources; only the very basic key principles and the notation are alike, cf. [Ape99] for *a priori* bounds and [Kun99] for *a posteriori* bounds. Since we are not primarily interested in *a priori* estimates, only some key references are given here, see [Ape99, RST96, MOS96, Lin03, HL98, FP01]. Our main interest lies in *a posteriori* error estimation.

1.2 Outline

A posteriori error estimation on anisotropic meshes has attracted some interest during the last decade, see [Sie96, DGP99, Kun99, Kun00, KV00, Kun01a, Kun01e, FPZ01, RL01, Gro02, Pic03, Ran01]. The earlier of these papers mainly establish the foundations and principles of anisotropic error estimation and provide key ingredients for their analysis. For example the comparatively simple Poisson problem is considered in [Sie96, Kun99,

DGP99]. Later on the interest turned to more challenging problems such as singularly perturbed equations [FPZ01, Kun01e, Kun01f, Kun03, Gro02], or the Stokes equations [CKN03, Ran01].

Our recent work shows a similar division. The Poisson problem as well as basic tools and principles for anisotropic error estimation have already been described and investigated in [Kun99]; see this source for a comprehensive exposition. In contrast, the work that is presented in this book completely falls into the latter class of more challenging problems; see this section for a description.

Let us now picture the content of this book. The present **Chapter 1** first introduces anisotropic discretizations and error estimation. Right now each of the chapters is briefly pictured. Some words about the presentation of the results follow in Section 1.3.

Chapter 2 starts with general notation. Then basic resources are presented, which mainly concern (anisotropic) bubble functions and the alignment of an anisotropic mesh. This required alignment of an anisotropic function and the anisotropic mesh seems to be an inherent key feature that is hidden in all known anisotropic *a posteriori* error estimators.

As it will be explained in Section 1.3 below, the exposition is kept to a minimum, i.e. to those tools that are later required to present and understand the results. If one is interested in the ingredients and tools for the proofs and the analysis then one should consult the respective paper and, if necessary, the basic principles given in [Kun99]. Section 2.4 assists the reader in this task and provides a more extensive list of ingredients.

After these preliminary topics we now turn to a first class of PDEs. **Chapter 3** is devoted to a *singularly perturbed reaction diffusion model problem*. This type of problem often serves as a simplified model for more complex (and often nonlinear) problems, such as the chemical reaction of several species, or semiconductor device simulation, or phase separation of a binary mixture (e.g. molten alloys), for example. Further interest in this PDE is stimulated by the fact that the problem is rich enough to exhibit many features, technicalities, and difficulties of real-world problems; yet it is accessible by a comparatively simple analysis.

The singularly perturbed character of the PDE frequently gives rise to boundary layers. They exhibit strong directional features, i.e. the solution is anisotropic. Thus a corresponding, anisotropic discretization can be applied advantageously. Our main attention focuses on tight *a posteriori* error estimation for these anisotropic meshes.

The chapter comprises four of our papers. The presentation starts with an introduction and motivation. Section 3.2 then investigates whether the energy norm is appropriate to measure the error. The material is taken from [Kun02a]. Next, a residual error estimator for the energy norm is proposed in Section 3.3. The results originate from [Kun01e]. A local problem error estimator for the energy norm can be found in Section 3.4; here the source is [Kun01f]. Two error estimators for the H^1 seminorm are presented in Section 3.5. Both of them are contained in [Kun01c]. The numerical experiments for all estimators are combined in Section 3.6.

There exist related papers. Most notably, the *reaction diffusion* problem considered here is a special case of the *convection diffusion* problem of Chapter 4 when the convection vanishes. Hence the papers mentioned there (and the results achieved) often apply here as well; see Chapter 4 for details. Nevertheless a separate presentation of both PDEs is beneficial since the reaction diffusion problem is considerably easier to treat (due to the vanishing convection). Apart from these convection related papers, we mention [Gro02] which covers a reaction diffusion problem as well. The achievements are partly based on our results. Finally we remark that short and preliminary considerations for a singularly perturbed model problem are already contained in [Kun99]. There a residual error estimator for the energy norm has been considered. Later the results have been corrected (concerning some basic inequality), the residual weights have been improved, Neumann boundary conditions have been included, and numerical experiments have been carried out.

Let us now briefly address *optimality* of our results. The answers are somewhat diverse. Starting with *energy norm* estimates, we achieve uniformity with respect to the small perturbation parameter (as in the isotropic case). With respect to the anisotropy, the estimates are uniform for the lower error bound but suboptimal for the upper error bound. For the latter bound one requires suitable anisotropic meshes to guarantee tight estimates (see Sections 2.3 and 3.3 for details).

For the H^1 seminorm the results are less fortunate. As above, one achieves optimality w.r.t the perturbation parameter, and suboptimality w.r.t. the anisotropy for the upper error bound. The lower error bound now also becomes suboptimal since it contains an additional L^2 norm term. It arises basically from the relation between the energy norm and the H^1 seminorm.

In **Chapter 4** a *singularly perturbed convection diffusion problem* is considered. Many physical applications lead to this type of PDE: Pollution propagation; groundwater transport; or semiconductor modelling, to name but a few. Furthermore, convection problems are often embedded in more complex models.

We are interested in the case of dominating convection which implies a much more complex behaviour of the exact solution u. For example, there may be boundary or interior layers, and there are different characterizations of the layer type. An appropriate discretization that reflects the layer features of the solution will contain anisotropic elements. Again we are mainly interested in *a posteriori* error bounds that are tight on anisotropic meshes.

In Section 4.1 the problem is stated first. Then we present two *a posteriori* error estimators. Both are described in [Kun03], and both aim at the error in an energy-like norm. The residual error estimator is given in Section 4.2 while a local problem error estimator is presented in Section 4.3. Several numerical examples of Section 4.4 complement and confirm the theory.

To our knowledge there is only one other paper [FPZ01] that covers *a posteriori* error estimation on anisotropic discretizations (for this PDE). The authors bound some *functional* of the error (i.e. they do not consider an error *norm*). To this end they utilize the corresponding dual solution, *a priori* interpolation estimates, and a postprocessing procedure. Both approaches [FPZ01] and [Kun03] cannot be compared directly because of the different aim and analysis.

The question of *optimality* of the results is more delicate than for the reaction diffusion problem of the previous chapter. Most notably, the lower and upper error bound may differ by a factor that is essentially determined by a so-called local mesh Peclet number. This local Peclet number relates the local convection, the diffusion, and the local mesh size. Tight error bounds can be guaranteed if this local Peclet number is small, i.e. $\mathcal{O}(1)$ at most. However, this somewhat unsatisfactory result does not arise from the anisotropic discretization but stems solely from the governing PDE. In other words, isotropic estimators face the same difficulty.

The optimality with respect to the anisotropy of the mesh can be answered similarly as above: The lower error bound holds uniformly while the upper error bound requires suitably aligned anisotropic meshes. Some of the effects related to optimality can be seen in the numerical experiments.

Chapter 5 is devoted to the Stokes equations which describe the movement of an incompressible viscous fluid. The Stokes problem can be obtained e.g. when simplifying and linearising the Navier Stokes equations. There one assumes a large viscosity and a negligible convection.

The Navier Stokes problem frequently induces solutions with layers which justify anisotropic discretizations. Therefore one is interested in suitable anisotropic *a posteriori* error estimators. Naturally many investigations start with the simplified Stokes problem.

Apart from this motivation, the Stokes problem itself is already interesting as far as anisotropy is concerned. In particular when a 3D domain with a concave edge is considered, one generically obtains a singular solution that exhibits anisotropic behaviour along that edge. Again anisotropic discretizations are appropriate [ANS01b, SSS99], and suitable *a posteriori* error estimators are sought.

As in the previous two chapters, the main focus is on such *a posteriori* error estimators that provide tight error bounds on anisotropic meshes. The material is taken from [CKN03] which comprises our joint work with S. Nicaise and E. Creusé. In that work we derive residual error estimators for a large variety of settings:

- 2D and 3D problems are considered,
- conforming and nonconforming methods are covered,
- many elements can be treated (triangles or rectangles in 2D, and tetrahedra, pentahedra, or hexahedra in 3D).

The exposition of the material starts with an introduction in Section 5.1. The problem is formulated and discretized by means of a mixed method in Section 5.2. Some additional notation is introduced there as well. Section 5.3 presents several examples of finite element pairs that fall within the scope of our approach. They are included here because of some side conditions which do not arise from the error estimation but (mainly) from the mixed method. Eventually Section 5.4 is devoted to the residual error estimators. First the lower error bound is stated. The subsequent upper error bound is split in the results for conforming and nonconforming discretizations. The numerical experiments of Section 5.5 accompany the theory.

As far as we know there is only the work [Ran01] which also addresses *a posteriori* error estimation on anisotropic meshes for the Stokes equations. There a nonconforming Crouzeix-Raviart discretization of the 2D Stokes problem is treated. A hierarchical estimator is proposed which requires a (comparatively strong) saturation assumption, as it is standard for this type of estimator. Unfortunately it is not clear whether Lemma 1 of [Ran01] can be derived as easily as claimed. In comparison, our approach is much more comprehensive and covers more settings.

The *optimality* of our results varies depending on the precise circumstances. Thus also the analysis is partially more elaborate than before. Let us start with the *lower error bound*. An optimal local lower bound is obtained for 2D meshes, and for conforming 3D discretizations, and for nonconforming 3D discretizations consisting of tetrahedra. The remaining case is characterized by 3D nonconforming discretizations containing pentahedra or hexahedra (i.e. there are rectangular faces). There only a *global* lower bound is achieved (and no local bound). We believe this result to be suboptimal.

Turn now to the *upper error bounds*. They are less favourable than before since the mesh anisotropy enters in a quite obscure way. For details we refer to the exposition of Chapter 5 and to [CKN03]. It is conceivable that our techniques to derive error estimates now reach certain limits. This opinion should of course considered with due care. The numerical experiments gave promising results, for example.

Finally the topic of **Chapter 6** is not about defining an error estimator as such. Instead it focuses on the interplay between error estimation and mesh construction, in particular in the framework of adaptive anisotropic algorithms (see Section 1.1 above). Two approaches lead to this chapter.

Firstly, consider the upper error bounds that are presented in Chapters 3–5. They are not completely uniform with respect to the mesh anisotropy since they contain a so-called *alignment measure*. This alignment measure specifies the influence of the anisotropic mesh on the error bound. The worse (i.e. larger) the alignment measure is, the worse and less tight the upper error bound becomes.

Unfortunately the alignment measure is unknown for the upper error bound since it contains the unknown exact solution u. Thus the question arises whether it is possible to construct suitably aligned anisotropic meshes which *automatically guarantee* a favourable (i.e. small) alignment measure. With some heuristic assumptions, the answer will be *Yes*.

For the second approach to the topic of this chapter, recall the anisotropic adaptive algorithm described above on page 4. Consider in particular the *information extraction* of step 3. At present, none of the existing anisotropic error estimators can provide all of the required anisotropic information. Instead this information is supplied by some external procedure. The so-called *Hessian strategy* is one such procedure that is frequently applied and used to construct anisotropic meshes, see also the references in Chapter 6.

It is not immediately clear if meshes constructed in this way are also suitable for anisotropic error estimation. Our investigation answers this question to the positive. Again some heuristic assumptions will be posed, since the Hessian strategy itself is heuristic in nature.

The value of our investigations mainly lies on the theoretical side, at least at a first glance:

- The investigations provide a theoretical justification of the Hessian strategy.
- Even more important, they justify the aforementioned *alignment measure*. Although this alignment measure is not known in practical applications, it will be favourable (i.e. small) on anisotropic meshes constructed via the Hessian strategy. This observation is now also of *practical importance*, since effectively one does not need to worry (too much) about the alignment measure. In other words, robustness of the anisotropic error estimation with respect to the mesh anisotropy is achieved for suitable anisotropic meshes in practice.

The questions considered here have not been addressed elsewhere in such a way. Furthermore the whole topic is closely connected to an observation which we believe to be an inherent key feature of anisotropic error estimation. Up to now, all anisotropic error estimators suffer from certain restrictions which can be interpreted in two ways.

- No anisotropic error estimator provides tight upper *and* lower error bounds (with respect to the mesh anisotropy) at the same time. Tight bounds require an anisotropic mesh that is suitably aligned with the anisotropic solution. In other words, the discretization has to be able to resolve anisotropic key features of the problem appropriately. This condition may be posed explicitly (e.g. via the alignment measure) or implicitly. For more details see Section 2.3.
- Alternatively, if one seeks error bounds from above and below that are totally independent of the mesh anisotropy then one would have to use *two* estimators, i.e. one for each bound. Both bounds then may differ substantially, that is they would be mathematically correct and independent of the mesh anisotropy but at the same time useless for practical applications.

Please note that this comment in not a proven fact; instead it is a personal impression based on many discussions with collaborators and the intense study of the topic in recent years.

1.3   Presentation of the results

This book comprises several papers that form a Habilitation thesis. Thus at the beginning of each chapter the underlying publications are stated, and the connection to the other material is explained. The following material is collected in this book.

- [Kun02a] investigates the suitability of the energy norm for a singularly perturbed reaction diffusion problem.
- [Kun01e] presents a residual error estimator for the energy norm for the same PDE.
- [Kun01f] gives a local problem error estimator for the energy norm for the same PDE.
- [Kun01c] is a slightly revised version of the technical report [Kun01d]. Currently it is under review. The work is devoted to H^1 seminorm error estimation for a singularly perturbed reaction diffusion problem.
- [Kun03] proposes and analyses two energy norm error estimators for convection dominated problems.
- [CKN03] is the most recent technical report. This is the only joint work; it has been submitted to a journal. Several residual error estimators for the Stokes problem are derived.
- [Kun02b] investigates the interplay of anisotropic error estimation and mesh construction.

All the material is already published, either as a journal article or a technical report. Therefore in this book we summarise the main results and all definitions that are necessary for it. We explain the philosophy and the key ideas behind our work. For example in Chapters 3–5 we motivate the problem, define the error estimators and present the error bounds. We comment on novel ideas and related work as well as the particular technicalities.

The proofs and technical tools are *not presented*. They can be found in the cited material, and they are not immediately necessary to understand the estimator or the error bound. However, the *key ideas of the proofs* are given. In this way the reader gains valuable insight without having to worry about details. When appropriate, severe technicalities are mentioned.

The overall intention of this style of presentation is to display the results in a compact and easily understandable way. We hope that practitioners in particular will find this helpful. Further and deeper insight is readily possible via the references.

The notation follows closely the original publications. Occasionally minor changes have been carried out to get a unified presentation in this book. For example, this concerns the squeezing parameter of equation (3.13) that is denoted by γ_E here.

Finally we remark that the motivation and the analysis of anisotropic error estimators generally allows two approaches. Firstly, one may start from *isotropic* estimators for that particular PDE and consider the necessary changes on anisotropic discretizations. Alternatively, one can recall anisotropic error estimators for several PDEs and then derive useful ideas for the problem under investigation. The reader will recognise both approaches in Chapters 3–5.

Chapter 2

Notation and basic resources

Here we introduce the notation and basic tools of the analysis. Hence this material can be found in all papers. The relation to the following chapters is obvious.

2.1 General notation

Let $\mathbb{P}^k$ and $\mathbb{Q}^k$ be the space of polynomials of total and partial degree not greater than k, respectively.

In order to avoid excessive use of constants, the shorthand notation $x \lesssim y$ and $x \sim y$ stands for $x \leq cy$ and $c_1 y \leq x \leq c_2 y$, respectively, with positive constants c, c_1, c_2. The constants do not depend on the terms x, y, or any discretization parameters or problem parameters.

Our exposition addresses problems in three dimensional as well as two dimensional domains. As a general rule, the more technical three dimensional case is described. When the two dimensional case is substantially different, it will be treated as well.

Domain, spaces, norms:
All problems will be posed on a open, bounded domain $\Omega \subset \mathbb{R}^d$, $d = 2$ or 3. The boundary $\partial\Omega$ is supposed to be polyhedral and consists of a (nonempty) Dirichlet part Γ_D and a Neumann part Γ_N (which may be empty).

For some domain ω, denote by $L^2(\omega)$ the standard Sobolev space of square integrable functions, equipped with the usual norm $\|\cdot\|_\omega$, cf. [Ada75]. The corresponding scalar product is $(v, w)_\omega = \int_\omega vw$. For the whole domain $\omega = \Omega$, the index of the norm and the scalar product will be omitted.

The Sobolev space of functions whose first derivatives are all in $L^2(\omega)$ is denoted by $H^1(\omega)$. By $H_0^1(\Omega)$ we mean that subspace of $H^1(\Omega)$ whose functions have vanishing trace on the Dirichlet boundary part Γ_D.

In Chapter 4 we require the space $L_\infty(\omega)$ of functions whose maximum norm $\|\cdot\|_{\infty,\omega}$ is bounded. There some problem data are measured in the maximum norm. Otherwise all norms are $L^2(\omega)$ norms, where the index ω denotes the domain.

For some vector $\mathbf{p}$ denote its Euclidean length by $|\mathbf{p}|$.

Triangulation and tetrahedron: In order to discretize the domain Ω, consider a family $\mathcal{F} = \{\mathcal{T}_h\}$ of triangulations $\mathcal{T}_h$ (also called *mesh*). For historical reasons we keep the single index h although a multi-index (h_1, h_2, h_3) would be more appropriate. Some mild conditions on the mesh are listed below.

Apart from Chapter 5, the meshes consist exclusively of tetrahedral elements (3D case) or triangular elements (2D). We start with a notation of these elements.

The four vertices of an arbitrary tetrahedron $T \in \mathcal{T}_h$ are denoted by $P_0, \dots, P_3$ such that P_0P_1 is the longest edge of T, $\text{meas}_2(\triangle P_0P_1P_2) \geq \text{meas}_2(\triangle P_0P_1P_3)$, and $\text{meas}_1(P_1P_2) \geq \text{meas}_1(P_0P_2)$.

In order to describe the main *anisotropic directions* of the tetrahedron, introduce three pairwise orthogonal vectors $\mathbf{p}_i$, see Figure 2.1.1. For the 2D case the notation is exactly as for the triangle $P_0P_1P_2$ in Figure 2.1.1.

The anisotropy vectors $\mathbf{p}_i$ are collected columnwise to form a matrix C_T given by

$$\left.\begin{array}{rcll} C_T & := & [\mathbf{p}_{1,T}, \mathbf{p}_{2,T}] \in \mathbb{R}^{2\times 2} & \text{in 2D} \\ C_T & := & [\mathbf{p}_{1,T}, \mathbf{p}_{2,T}, \mathbf{p}_{3,T}] \in \mathbb{R}^{3\times 3} & \text{in 3D.} \end{array}\right\} \tag{2.1}$$

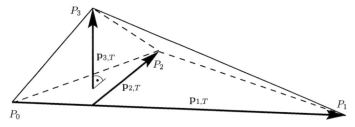

Figure 2.1.1: Notation of tetrahedron T

The *anisotropic lengths* of an element are given by

$$h_{i,T} := |\mathbf{p}_i|. \tag{2.2}$$

Observe $h_{1,T} > h_{2,T} \geq h_{3,T}$.

The minimal anisotropic length of a tetrahedron/triangle is of particular importance throughout the analysis. For a unified notation define thus

$$h_{min,T} := \min_{i=1\dots d} = h_{d,T} = \begin{cases} h_{2,T} & \text{for triangle (2D case),} \\ h_{3,T} & \text{for tetrahedron (3D case).} \end{cases} \tag{2.3}$$

The ratio of the maximal and minimal anisotropic length is termed *aspect ratio* or *stretching ratio*.

Elements are denoted by T, T_i or T'. Faces of a tetrahedron (3D case) and edges of a triangle (2D) are denoted by E or E_i. To shorten the notation, the measure of these items is given by

$$|T| := \text{meas}_d(T) \qquad |E| := \text{meas}_{d-1}(E).$$

Furthermore we require the length of the *height* over a face E (in a tetrahedron T, 3D case) or over an edge E (in a triangle T). Hence set

$$h_{E,T} := d \cdot \frac{|T|}{|E|}. \tag{2.4}$$

Normal vectors and jumps: Gradient jumps across interelement faces E (or edges in 2D) play an important role when defining error estimators.

For a face (or edge in 2D) of an element T the outer unitary normal vector n is defined as usual. Furthermore, for each face E we fix one of the two normal vectors and denote it by n_E.

The *jump* of some (scalar or vector valued) function v across a face E at a point $y \in E$ is then defined as

$$[\![v(y)]\!]_E := \begin{cases} \lim_{t \to +0} v(y + tn_E) - v(y - tn_E) & \text{for an interior face } E, \\ v(y) & \text{for a boundary face } E. \end{cases}$$

Note that the sign of $[\![v]\!]_E$ depends on the orientation of n_E. However, terms such as a gradient jump $[\![\nabla v \, n_E]\!]_E$ are independent of this orientation.

Auxiliary subdomains: Let $T \in \mathcal{T}_h$ be an arbitrary tetrahedron. Following [Ver96], let ω_T be that domain (or patch) that is formed by T and all tetrahedra that have a common face with T. Note that ω_T consists of less than five tetrahedra if T has a boundary face.

Let E be an inner face (triangle) of $\mathcal{T}_h$, i.e. there are two tetrahedra T_1 and T_2 having the common face E. Set the domain $\omega_E := T_1 \cup T_2$. If E is a boundary face set $\omega_E := T$ with $T \supset E$.

Mesh requirements: The mesh has to satisfy some mild standard assumptions.

1. The mesh is conforming in the standard sense of [Cia78, Chapter 2].
2. A node x_j of the mesh is contained only in a bounded number of elements.
3. The size of neighbouring elements does not change rapidly, i.e.

$$h_{i,T_1} \sim h_{i,T_2} \qquad \forall i = 1 \dots d, \forall T_1 \cap T_2 \neq \emptyset.$$

Remark 2.1.1 In certain situations we do not want to use *element based* quantities (such as $h_{min,T}$) but utilize *face related* terms instead. For example consider an interior face $E = T_1 \cap T_2$, and define the terms

$$h_E := (h_{E,T_1} + h_{E,T_2})/2 \qquad , \qquad h_{min,E} := (h_{min,T_1} + h_{min,T_2})/2 \quad .$$

Their advantage is that they are no longer related to T_1 or T_2 but to E. They clearly satisfy $h_E \sim h_{E,T_i}$ and $h_{min,E} \sim h_{min,T_i}$. For a boundary face $E \subset \partial T \cap \Gamma$ define similarly $h_E := h_{E,T}$ and $h_{min,E} := h_{min,T}$. Similar to above one can infer $h_{min,E} < 2h_E$ for all faces E. $\qquad\qquad \square$

2.2 Bubble functions

Another useful and important tool are so-called bubble functions which are applied, for example, for defining the local problem and its ansatz space but also for the analysis. The bubble functions were already partially introduced in [Ver96] and [Kun01e].

Denote by $\lambda_{T,1}, \cdots, \lambda_{T,4}$ the barycentric coordinates of an arbitrary tetrahedron T. The *element bubble function* b_T is defined by

$$b_T := 4^4 \cdot \lambda_{T,1} \cdot \lambda_{T,2} \cdot \lambda_{T,3} \cdot \lambda_{T,4} \in \mathbb{P}^4(T) \qquad \text{on } T \quad . \tag{2.5}$$

For simplicity assume that b_T is extended by zero outside its original domain of definition.

Further we require face bubble functions. To this end let $E = T_1 \cap T_2$ be an inner face (triangle) of $\mathcal{T}_h$. Enumerate the vertices of T_1 and T_2 such that the vertices of E are numbered first, and introduce the functions

$$b_{E,T_i} := 3^3 \cdot \lambda_{T_i,1} \cdot \lambda_{T_i,2} \cdot \lambda_{T_i,3} \qquad \text{on } T_i, \ i = 1, 2 \quad .$$

The *standard face bubble function* $b_E \in C^0(\omega_E)$ is now defined in a piecewise fashion (with support $\omega_E = T_1 \cup T_2$) by

$$b_E := \begin{cases} b_{E,T_1} & \text{on } T_1 \\ b_{E,T_2} & \text{on } T_2 \\ 0 & \text{otherwise} \end{cases}, \qquad (2.6)$$

see also the middle of Figure 2.2.3.

The bubble functions from above suffice to analyse the error estimator for the *Poisson* equation, cf. [Ver96, Kun01a]. However, for *singularly perturbed problems* as considered in Chapters 3 and 4 we have to introduce modified face bubble functions, cf. also [Kun01e, Ver98b]. To this end a so-called squeezed tetrahedron $T_{E,\gamma}$ will be set up.

Squeezed tetrahedron $T_{E,\gamma}$: The concept of the squeezed tetrahedron has been introduced in [Kun01e] and originates from [Ver98b] (in a simpler, modified form there). Here we only repeat the definition; further details and auxiliary results can be found in [Kun01e].

Start with an arbitrary but fixed tetrahedron T. The *squeezed tetrahedron* $T_{E,\gamma} \subset T$ is a sub–tetrahedron of T which depends on a face E of T and a real number $\gamma \in (0, 1]$. For its precise definition, enumerate the vertices of T temporarily such that $E = Q_1 Q_2 Q_3$ and $T = O Q_1 Q_2 Q_3$, cf. Figure 2.2.2. Introduce barycentric coordinates such that λ_0 is related to O, and $\lambda_1, \lambda_2, \lambda_3$ correspond to Q_1, Q_2, Q_3, respectively.

Let P be that point with barycentric coordinates

$$\lambda_0(P) = \gamma \qquad \text{and} \qquad \lambda_1(P) = \lambda_2(P) = \lambda_3(P) = \frac{1-\gamma}{3} \quad .$$

Then let $T_{E,\gamma}$ be the tetrahedron with vertices P and Q_1, Q_2, Q_3, i.e. $T_{E,\gamma}$ has the same face E as T but the fourth vertex is moved towards E with the rate γ.

An alternative description is as follows. With S_E being the midpoint (i.e. center of gravity) of face E, point P lies on the line $S_E O$ such that $|S_E P| = \gamma \cdot |S_E O|$. Note that for $\gamma = 1$ one gets $T_{E,\gamma} \equiv T$ whereas in the limiting case $\gamma \to 0$ the tetrahedron $T_{E,\gamma}$ collapses to the face E.

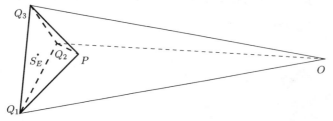

Figure 2.2.2: Tetrahedra $T = O Q_1 Q_2 Q_3$ and $T_{E,\gamma} = P Q_1 Q_2 Q_3$

Squeezed face bubble function $b_{E,\gamma}$: Now we are ready to define the aforementioned squeezed face bubble function. In order to avoid the precise but somewhat technical definition of [Kun01e], we proceed here with a verbal description combined with some figures (for the 2D case).

Start with some face E and let T_1, T_2 be its two neighbouring tetrahedra, i.e. $\omega_E = T_1 \cup T_2$. For an arbitrary real number $\gamma \in (0,1]$ consider both squeezed tetrahedra $T_{1,E,\gamma} \subset T_1$ and $T_{2,E,\gamma} \subset T_2$, cf. Figures 2.2.2 and 2.2.3 (top).

The so–called *squeezed face bubble function* $b_{E,\gamma}$ acts only on $T_{1,E,\gamma} \cup T_{2,E,\gamma} \subset \omega_E$. Its piecewise definition is

$$b_{E,\gamma} := \begin{cases} b_{E,T_{1,E,\gamma}} & \text{on } T_{1,E,\gamma} \\ b_{E,T_{2,E,\gamma}} & \text{on } T_{2,E,\gamma} \\ 0 & \text{on } \omega_E \setminus (T_{1,E,\gamma} \cup T_{2,E,\gamma}) \end{cases} \tag{2.7}$$

cf. the bottom of Figure 2.2.3. Note that the squeezed face bubble function on T_i can be viewed equivalently as the standard face bubble function on the squeezed tetrahedron $T_{i,E,\gamma}$, i.e.

$$b_{E,\gamma}\big|_{T_i} \equiv b_{E,T_{i,E,\gamma}} \quad .$$

Figure 2.2.3 may facilitate the understanding of the standard/squeezed face bubble function for the two–dimensional case. For boundary faces one restricts $b_{E,\gamma}$ to the unique tetrahedron with $\partial T \supset E$.

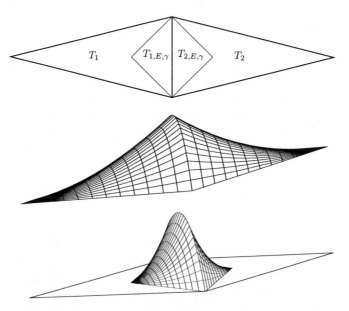

Figure 2.2.3: Top: ω_E and squeezed triangles $T_{i,E,\gamma}$ (2D case)
Middle: standard face bubble function b_E
Bottom: squeezed face bubble function $b_{E,\gamma}$

2.3   Alignment measure

In order to motivate the alignment measure properly, let us start with an important difference between error estimation on *isotropic* and *anisotropic* meshes. For isotropic meshes
the error estimation is valid no matter what actual mesh is used [Ver96, AO00]. In contrast, this feature is lost on anisotropic meshes where all known error estimators require the
anisotropy of the mesh to be aligned with the anisotropy of the solution. Heuristically this
means that anisotropic elements (e.g. tetrahedra) are stretched in that direction where the
solution shows little variation. If this requirement is violated then the upper and lower error
bound may differ by a large factor (depending on the degree of the misalignment of mesh
and solution). This gap may be as large as the maximum stretching ratio of all elements of
the mesh which would render the error bounds useless.

In order to investigate this matter mathematically, let us recall the proposals from known
(analytically based) anisotropic error estimators. Siebert [Sie96] restricts the set of treatable
anisotropic functions. In [Kun99, KV00, Kun00, Kun01e, Gro02] a so–called matching
function $m_1(v, \mathcal{T}_h)$ is introduced that measures the alignment of an anisotropic function v
and an anisotropic mesh $\mathcal{T}_h$. In [DGP99] and [Ran01] certain saturation assumptions are
necessary that imply a similar correspondence of the anisotropic mesh and the anisotropic
solution. Post-processing based estimators as proposed e.g. by [Pic03] require a sufficiently
accurate solution u_h if post-processing is to improve u_h. Again this can only be expected for
suitably aligned anisotropic meshes. The same holds true for estimators for error functionals
(which employ a dual solution, cf. [FP00]). Summarising, all known error estimators on
anisotropic discretizations inherit certain restrictions on the anisotropic mesh which may
be formulated explicitly or implicitly.

In our work the alignment of the anisotropic mesh and the anisotropic function will
be measured explicitly. To this end the *alignment measure* will be proposed. Historically
it has been introduced as *matching function* [Kun99, Kun00] but later renamed *alignment
measure*. Here it will be utilized in Chapters 3, 4 and 6. For the Stokes problem of Chapter 5
a slightly modified notation will be used. The definition now reads as follows.

Definition 2.3.1 (Alignment measure; previously called Matching function)
*Let $v \in H^1(\Omega)$, and $\mathcal{T}_h \in \mathcal{F}$ be a triangulation of Ω. Define the alignment measure
$m_1 : H^1(\Omega) \times \mathcal{F} \mapsto \mathbb{R}$ by*

$$m_1(v, \mathcal{T}_h) := \left(\sum_{T \in \mathcal{T}_h} h_{min,T}^{-2} \cdot \|C_T^\top \nabla v\|_T^2 \right)^{1/2} \bigg/ \|\nabla v\| \quad . \tag{2.8}$$

Note that the entries of the vector $C_T^\top \nabla v \equiv (\mathbf{p}_1^\top \nabla v, \mathbf{p}_2^\top \nabla v, \mathbf{p}_3^\top \nabla v)^\top$ can also be viewed as
scaled directional derivatives along the orthogonal directions $\mathbf{p}_i$ (recall $|\mathbf{p}_i| = h_{i,T}$).

To deepen the understanding of the alignment measure let us briefly discuss its behaviour
and influence. More details and a comprehensive discussion can be found in [Kun99, Kun00].

By defining temporarily $h_{max,T} := h_{1,T}$, one obtains

$$1 \leq m_1(v, \mathcal{T}_h) \lesssim \max_{T \in \mathcal{T}_h} \frac{h_{max,T}}{h_{min,T}} \quad .$$

Although this crude upper bound is useless for practical purposes it implies $m_1 \sim 1$ on
isotropic meshes. Then m_1 merges with other constants and becomes invisible; in this sense
(2.8) is an extension of the theory for isotropic meshes. If an anisotropic mesh $\mathcal{T}_h$ is *well
aligned* with an anisotropic function v then one also obtains $m_1(v, \mathcal{T}_h) \sim 1$. If, however,

the anisotropic meshes are *not aligned* with the function then the alignment measure can be as large as the maximal aspect ratio of the elements of the mesh, i.e. $m_1(v, \mathcal{T}_h) \gg 1$. Numerical experiments with well aligned meshes result in values in the range of $2 \ldots 4$, cf. Section 3.6 for example.

The influence of the alignment measure m_1 can be seen in the error bounds of Theorem 3.3.4, 3.4.4, 3.5.2, 3.5.5, 4.2.4 and 4.3.4. Roughly speaking, tight upper and lower error bounds can only be expected if the alignment measure is small, i.e. $m_1(u - u_h, \mathcal{T}_h) \sim 1$. This can be obtained with an appropriate anisotropic mesh. Chapter 6 is devoted to all these issues in more detail.

For practical applications it is important to realize that the aforementioned error bounds involve the alignment measure $m_1(u - u_h, \mathcal{T}_h)$ for the *unknown error* $u - u_h$. In this sense the error bounds are unusual since they contain a term that cannot be evaluated explicitly. This feature seems to be quite a drawback at first sight. A closed inspection weakens this disadvantage quite considerably.

- Firstly, all known anisotropic error estimators inherit similar restrictions. The alignment measure is one (readily conspicuous) possibility to formulate such a restriction.
- Appropriate anisotropic meshes yield a small alignment measure $m_1(u - u_h, \mathcal{T}_h) \sim 1$, both in practice and theoretically, cf. Chapter 6.
- The alignment measure can be approximated. To this end recall (2.8) and realize that $m_1(u - u_h, \mathcal{T}_h)$ contains ∇u but not u itself. Hence one may replace the exact gradient ∇u by some *recovered gradient* $\nabla^R u_h$ to obtain an approximation $m_1^R(u_h, \mathcal{T}_h) \approx m_1(u - u_h, \mathcal{T}_h)$. Numerical experiments show quite a good agreement, see e.g [Kun00, Kun01e].
- The alignment measure is necessary for a rigorous analysis. In practice, ons will take care of appropriate anisotropic meshes. Then one may simply omit the alignment measure (or its approximation), as it will be small.

2.4 Useful tools

Before some useful tools are presented, we recall the scope of the whole work. Our main intention is to provide an overview about recent achievements on anisotropic error estimation. The exposition should allow easy access to the main results (and the required definitions) such that an implementation is possible without having to cope with much theoretical overhead. Hence we explain the methodology and key ideas of the proofs but omit the actual proofs and technical resources. The literature provides enough means to fill this gap.

Nevertheless a brief list of useful tools may be advantageous to get an impression of the techniques and resources that are used. All of them are carefully tailored to cope with the anisotropy of the discretization. The transformation technique is always a key ingredient, i.e. anisotropic elements are transformed to reference elements to remove the influence of the anisotropy.

Upper error bound: The main ingredients are always specific *anisotropic interpolation error estimates*. These estimates depend on the actual PDE. Furthermore there is a close relationship to the definition of the (anticipated) error estimator and error bound. This can be seen exemplarily on [Kun01e, Lemma 3.10] and [CKN03, Lemma 4.4]. Furthermore the proof of the interpolation estimates partly requires an *anisotropic trace theorem*, cf. [Kun01e, Lemma 3.5].

Lower error bound: Basic ingredients are always *bubble functions*. For singularly perturbed problems they have to be modified which leads to the *squeezed face bubble func-*

tions of Section 2.2. Closely related are *inverse inequalities* for these bubble functions, see e.g. [Kun01e, Lemmas 3.6 and 3.7], [Kun03, Lemma 2] or [CKN03, Lemma 4.1].

The analysis of local problem error estimators requires certain *inequalities for the (finite dimensional) local problem space*, see e.g. Lemma 3.4.2. These inequalities are particularly technical to prove because of the great variability of the local space. This is underlined by [Kun01f, Appendix A].

Chapter 3

Singularly perturbed reaction diffusion problem

The material of this chapter is contained in [Kun02a, Kun01e, Kun01f, Kun01c]. The common feature with the material from other chapters is mainly given by the *anisotropy* of the solution. This in turn leads to similar approaches of the analysis which often requires similar tools and requisites, cf. Chapter 2.

3.1 Problem description

Consider a singularly perturbed reaction diffusion model problem with Dirichlet–Neumann boundary conditions

$$\left. \begin{aligned} -\varepsilon \Delta u + u &= f \quad \text{in } \Omega \\ u &= 0 \quad \text{on } \Gamma_{\mathrm{D}} \\ \varepsilon \cdot \partial u / \partial n &= g \quad \text{on } \Gamma_{\mathrm{N}} \end{aligned} \right\} \tag{3.1}$$

in a bounded, polyhedral domain $\Omega \subset \mathbb{R}^d$, $d = 2, 3$, with boundary $\partial \Omega = \Gamma_{\mathrm{D}} \cup \Gamma_{\mathrm{N}}$, $\Gamma_{\mathrm{D}} \cap \Gamma_{\mathrm{N}} = \emptyset$. The perturbation parameter $\varepsilon > 0$ can become arbitrarily small.

Assume $f \in L^2(\Omega)$, $g \in L^2(\Gamma_{\mathrm{N}})$ and $\mathrm{meas}_{d-1}(\Gamma_{\mathrm{D}}) > 0$. The corresponding variational formulation for (3.1) becomes:

$$\left. \begin{aligned} \text{Find } u \in H_0^1(\Omega) : \quad & a(u,v) = \langle F, v \rangle \qquad \forall v \in H_0^1(\Omega) \\ \text{with} \quad a(u,v) := & \int_\Omega \varepsilon \cdot (\nabla u)^\top \nabla v + u\,v \qquad \langle F, v \rangle := \int_\Omega f v + \int_{\Gamma_{\mathrm{N}}} g v \end{aligned} \right\} \tag{3.2}$$

We utilize a family $\mathcal{F} = \{\mathcal{T}_h\}$ of triangulations $\mathcal{T}_h$ of Ω. Let $V_{0,h} \subset H_0^1(\Omega)$ be the space of continuous, piecewise linear functions over $\mathcal{T}_h$ that vanish on Γ_{D}. Then the finite element solution $u_h \in V_{0,h}$ is uniquely defined by

$$a(u_h, v_h) = \langle F, v_h \rangle \qquad \forall v_h \in V_{0,h} . \tag{3.3}$$

Due to the Lax–Milgram Lemma both problems (3.2) and (3.3) admit unique solutions. Furthermore the bilinear form $a(\cdot, \cdot)$ naturally defines an energy and an *energy norm* by

$$\|v\|^2 := a(v,v) = \varepsilon \|\nabla v\|^2 + \|v\|^2. \tag{3.4}$$

Singularly perturbed reaction diffusion problems frequently lead to (exponential) boundary layers whenever the right-hand side f and the boundary data do not match. The smaller the perturbation parameter ε is, the more distinguished the layers will be, cf. also [RST96, MOS96, Mor96]. Nevertheless problem (3.2) enjoys several favourite properties:

19

- The problem is *symmetric*,
- it allows the definition of an *energy* and a related *norm*.
- The bilinear form $a(\cdot,\cdot)$ is elliptic and bounded (with respect to the energy norm) with constants independent of ε. More precisely, these constants are equal to 1.
- Consequently problem (3.3) can be solved numerically without stabilization (which is required e.g. for *convection dominated* problems, cf. Chapter 4).

Let us now consider the finite element method and some discretization aspects in particular. Standard methods employ so-called *isotropic* meshes. That is, the elements are shape regular or, equivalently, the ratio of the diameters of the circumscribed and inscribes spheres is bounded. However, the singularly perturbed problem stated above admits a solution with strong directional features such as boundary or interior layers. Hence it is natural to reflect the anisotropy of the solution u in the mesh $\mathcal{T}_h$. Such a mesh will have stretched, *anisotropic* elements, i.e. elements that are no longer shape regular. The figures of Section 3.6 provide a lively image of such anisotropic meshes. Furthermore we remark that hp finite element discretizations gained popularity for singular parturbations, see [Mel02] and its extensive bibliography.

Our main focus here is on *a posteriori error estimation*. This topic has been well analysed by now for *isotropic* discretizations of elliptic problems, cf. the standard textbooks [Ver96, AO00]. The situation is much less clear for singularly perturbed problems. Although we still have ellipticity in our case, the corresponding ellipticity constant deteriorates as the perturbation parameter ε tends to zero. As a consequence it is much more difficult to find lower and upper error bounds that are tight and *uniform w.r.t. ε*. For problem (3.1) this goal has been achieved within the last decade by [Ang95, Ver98b, AB99, KS01].

A second technicality of error estimation arises from the *anisotropy* of the discretization. In contrast to isotropic meshes, the error bounds may now depend on the maximum stretching ratio of the anisotropic elements of the mesh. Therefore there may be a large gap between the lower and upper error bound (or, equivalently, between the error and the error estimator). Thus there has been increasing research to derive error estimators which are uniform with respect to the mesh anisotropy (even if this goal is not completely achieved yet). First results can be attributed to [Sie96, Kun99, DGP99, FPZ01], to name but a few.

Following these considerations, we are interested in error estimation for the singularly perturbed reaction diffusion problem (3.2) with an anisotropic discretization. The main purpose (and technicality) of our analysis is to find lower and upper error bounds that are uniform in the small perturbation parameter ε, *and* that are tight with respect to the mesh anisotropy. This goal has not been achieved before, i.e. the results presented in our work here are new.

(Partially) based on our achievements, an error estimator based on the so-called equilibrated residual method has been proposed later in [Gro02]. It is an extension of the isotropic theory of [AB99]. The estimator is obtained by solving an infinite dimensional local problem which has to be approximated in practice.

Primarily we concentrate on the most natural norm related to (3.2), namely the *energy norm* introduced above. This norm has been used also by other authors [AB99, Ver98b, RST96]. Before we proceed with the analysis, however, we have to ask if this energy norm is suitable and sensible to choose. Following [Kun02a], Section 3.2 sheds light on this question. The answer is not a complete *Yes* but stresses the purpose of the computation instead.

In Section 3.3 we then present an error estimator for the energy norm which is based on evaluating the residuals. This estimator is reliable and efficient as well as robust with respect to the small perturbation parameter ε. The definitions and results are mainly taken from [Kun01e].

Another energy norm error estimator that is based on the solution of (small) local problems is presented in Section 3.4. Again this estimator is reliable, efficient and robust. The presentation originates from [Kun01f].

Finally in Section 3.5 we investigate error estimators in a different norm, namely the H^1 seminorm. It turns out that the estimators and the error bounds are mainly determined by the relation between the energy norm and the H^1 seminorm, cf. [Kun01d]. This results in somewhat less favourable error bounds than one might hope for.

The relation and the connection of all this material is clearly given by the same governing PDE. Two of the papers even aim at error bounds in the same norm.

3.2 Is the energy norm suitable?

The material of this section originates from [Kun02a].

3.2.1 Motivation

The question of a proper norm is fundamental for singularly perturbed problems and their analysis. In the literature, one often encounters the L_∞ or maximum norm (see e.g. [MOS96]), the energy norm [AB99, Ver98b, RST96, Ber02, XF03], or norms that are adapted to the solution method (e.g. the SDFEM norm in a slightly different context [RST96, ST01]).

Generally the maximum norm is somewhat stronger than the energy norm. In other words, a discrete solution with a small error in the maximum norm (usually) also implies a small error in the energy norm but not necessarily vice versa. Occasionally this observation leads to the (more or less informal) opinion that the maximum norm should be the preferred choice, and that the energy norm is useless. However, it is essential to note that the motivating question is incomplete - one has to ask: *Suitable for what?*

The aim of a numerical analyst should be to provide a solution with the accuracy required. It is important to realize that proper modelling of the underlying physical problem also includes the choice of an *appropriate norm* in which to measure the accuracy. This suggests that researchers interested in practical problems and adaptive methods may find the energy norm adequate for investigating singularly perturbed problems. If one is interested in problem independent norms and ε–uniform methods, the maximum norm can be the preferred choice.

We stress that both research aims and both choices of norms are well justified, depending on the underlying problem and on the modelling. Although most numerical analysts readily agree with this statement, we have often experienced that researchers use their preferred choice of norm but reject vigorously other norms. Hence we emphasize that different norms cannot be viewed as contradictory but they reflect instead the different modelling background and research purpose.

With this view in mind, the aim of this section is as follows:

> *Derive a simple adaptive algorithm that is based on a robust energy norm error estimator. Observe its numerical performance and conclude whether the algorithm and the energy norm are appropriate.*

The criterion to judge the performance of the algorithm is the error in the energy norm (or more precisely, the error decrease). We utilize a simple one dimensional (and hence isotropic) model problem. A successful adaptive algorithm is a prerequisite to carry out further research on anisotropic 2D/3D discretizations.

We emphasize that our focus is not primarily on the theory of error estimation or of adaptive algorithms. Instead we are essentially concerned about the *numerical performance* of the algorithm for practical examples. Apparently and surprisingly, such a numerical evaluation has not been carried out yet. The importance of these investigations is emphasized by the results of the numerical experiments. As it will be seen below, the adaptive algorithm (and the energy norm in particular) produce optimal meshes (w.r.t. the error decrease). This provides basic and important knowledge to justify subsequent research (see Sections 3.3 and 3.4 below). Because of the danger of misinterpretation we stress that our aim is neither to derive ε–uniform methods, nor to discourage or condemn the use of the L_∞ norm.

3.2.2 The adaptive algorithm

In order to concentrate on the role of the energy norm and to eliminate unwanted influences we retreat to a comparatively simple 1D model problem:

$$-\varepsilon\,u'' + u = f \quad \text{in } \Omega = (0,1), \qquad u(0) = 1, \quad u(1) = 0 \qquad (3.5)$$

with $f \in L^2(\Omega)$. It enjoys several favourite properties, namely

- it is a one–dimensional problem,
- the analytical solution u and thus the discretization error $u - u_h$ are known,
- the computational implementation is easily accomplished, e.g. in `MATLAB`, cf. [Kun01b].

The discretization with linear finite elements utilizes a mesh with nodal points $\{x_i\}_{i=0}^N$. The finite element solution is denoted by u_h. The adaptive algorithm consists of the steps *Solve system of equations – Estimate error – Refine mesh*. The last two ingredients are described now.

Error estimation: The energy norm residual error estimator is the one–dimensional counterpart of [Ver98b]. For a proper description, define temporarily the following data.

Mesh size	$h_i := x_i - x_{i-1}$	$i = 1 \ldots N$
Finite element	$T_i := (x_{i-1}, x_i)$	$i = 1 \ldots N$
Macro element	$\omega_i := (x_{i-2}, x_{i+1})$	$i = 2 \ldots N - 1$
	$\omega_1 := (x_0, x_2) \qquad \omega_N := (x_{N-2}, x_N)$	
Element residual	$R_i := f - (-\varepsilon\,u_h'' + u_h)$	$i = 1 \ldots N$
Jump residual	$J_i := \varepsilon \cdot [u_h'(x_i + 0) - u_h'(x_i - 0)]$	$i = 1 \ldots N - 1$
	$J_0 := J_N := 0$	
Scaling factor	$\alpha_i := \min\{1\,,\ \varepsilon^{-1/2} h_i\}$	$i = 1 \ldots N$
Local error estimator	$\eta_i^2 := \alpha_i^2\, \|R_i\|_{T_i}^2 + \varepsilon^{-1/2}\alpha_i \cdot (J_{i-1}^2 + J_i^2)$	$i = 1 \ldots N$

Here data approximation terms related to f are omitted (cf. [Ver98b]). This is justified when f is piecewise polynomial (of a fixed degree), as in our numerical examples. Note that x_i and J_i are node related data whereas h_i, T_i, R_i, α_i, η_i are element related data. Verfürth [Ver98b] has proven that the energy norm of the error is bounded locally from below and globally from above:

$$\eta_i \ \leq\ c_L\, \|\!|u - u_h|\!\|_{\omega_i} \qquad \forall\, i = 1 \ldots N,$$

$$\|\!|u - u_h|\!\|_\Omega \ \leq\ c_U \left(\sum_{i=1}^N \eta_i^2 \right)^{1/2}\ .$$

The constants c_L, c_U are independent of ε, i.e., the error estimation is robust.

Mesh refinement: Here it suffices to choose a simple strategy. Start with an equidistributed mesh of 10 elements. Once the error estimators are computed, an element T_i is bisected iff

$$\eta_i \geq \gamma \cdot \max_{k=1...N} \eta_k \quad ,$$

where the refinement parameter is set to $\gamma := 0.1$. More sophisticated refinement strategies are possible of course.

Optimality criterion: As described before, we judge the algorithm by the decrease of the error in the energy norm. The optimal rate is $\mathcal{O}(N^{-1})$.

3.2.3 Numerical experiments

After fruitful discussion with several researches a number of particularly interesting cases of problem (3.5) have been isolated and solved by the adaptive algorithm. Since all examples have yielded similar results, we discuss exemplarily the problem

$$-\varepsilon\, u'' + u = 2\varepsilon + x(1 - x) \quad \text{in } \Omega = (0, 1), \qquad u(0) = 1, \quad u(1) = 0.$$

The analytical solution behaves like $e^{-x/\sqrt{\varepsilon}} + x(1 - x)$ for $\varepsilon \ll 1$. It features a typical exponential boundary layer (at $x = 0$) which is superposed on a quadratic function, cf. Figure 3.2.1. This experiment investigates whether a layer can be detected in the presence of other errors. In the context of asymptotic expansions, the layer function can be seen as the inner expansion whereas the quadratic function represents (exemplarily) the outer expansion.

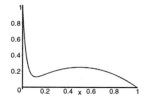

Figure 3.2.1: Analytic solution u for the numerical example, $\varepsilon = 10^{-3}$

In Figure 3.2.2 we present the results of our adaptive algorithm for different values of ε. The algorithm is judged by the *optimality criterion*, i.e. the energy norm error decrease with respect to the number of unknowns N. It turns out that in all cases the convergence rate is very close to the optimal rate of $\mathcal{O}(N^{-1})$ as soon as the boundary layer is resolved. As desired, the convergence rate is independent of ε. Hence the chosen algorithm is optimal since the best possible energy norm error decrease is attained. It is noteworthy that optimality is achieved although the energy norm of the layer function is $\mathcal{O}(\varepsilon^{1/4})$, and thus much smaller than the energy norm of the quadratic superposition (which is $\mathcal{O}(1)$).

Finally we remark that several other tests have been performed, e.g. with discontinuous data such as $f = \text{sign}(x - 0.5)$, or for problems with non–vanishing consistency error. The results are similar, cf. also [Kun01b].

3.2.4 Conclusions

With the help of numerical simulation one wants to solve (physical) problems up to a prescribed accuracy. The norm in which to measure this accuracy depends both on the research purpose and the modelling.

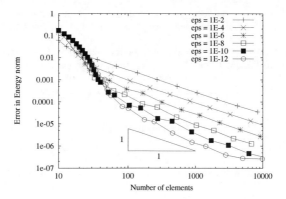

Figure 3.2.2: Error decrease in the energy norm

Here we have considered a singularly perturbed reaction diffusion model problem for which a robust adaptive algorithm has been sought and proposed. As a key ingredient for such methods one requires robust *a posteriori* error estimators which are known for the *energy norm*.

We have investigated the numerical performance of the proposed adaptive algorithm (this investigation has been lacking so far). The results have shown optimal behaviour of the adaptive method, and hence the energy norm is well suited within this framework. Our inquiries and its results also justify further research in more complicated situations, cf. the sections below.

3.3   Residual error estimator

The material of this section originates from [Kun01e]. Here the presentation of the results is slightly modified:

- Neumann boundary data are included,
- the approximation of the residuals is different.

Recall first the PDE (3.1) to be solved,

$$\left.\begin{array}{rcll} -\varepsilon\Delta u + u & = & f & \text{in } \Omega \\ u & = & 0 & \text{on } \Gamma_\mathrm{D} \\ \varepsilon \cdot \partial u/\partial n & = & g & \text{on } \Gamma_\mathrm{N} \end{array}\right\}$$

and the corresponding weak formulation (3.2). We aim at an error bound in the energy norm $\|\|v\|\|^2 \equiv \varepsilon\|\nabla v\|^2 + \|v\|^2$.

Residual error estimators bound the error $u - u_h$ by measuring the residual. However, instead of computing the norm of the residual in the dual space $[H_0^1(\Omega)]^* = H^{-1}(\Omega)$, one tries to obtain an equivalent measure by evaluating easier terms that involve the given data (e.g. f, Ω, or $\mathcal{T}_h$). The main task is to carefully calibrate the weights of the residual norms such that both an upper and lower error bound hold. The difficulties that arise from the singularly perturbed problem are here even emphasized and amplified by the anisotropic elements.

The methodology to obtain a lower error bound requires residual terms from a finite dimensional space [Ver96, Kun99]. Hence we replace the exact element residual by an approximate element residual which is constant over an element T (e.g. by means of an L^2 projection into $\mathbb{P}^0(T)$). Proceed analogously for the face residuals where g is replaced by g_h which is piecewise constant over the Neumann faces. The precise definitions are as follows.

Definition 3.3.1 (Element and face residual) *The exact element residual over an element T is given by*

$$R_T := f \, - \, (-\varepsilon \Delta u_h + u_h) \qquad on \ T.$$

The (approximate) *element residual r_T is any approximation to R_T that is constant on T, i.e.*

$$r_T \in \mathbb{P}^0(T). \tag{3.6}$$

For a face E define the (approximate) *face residual $r_E \in \mathbb{P}^0(E)$ by*

$$r_E := \begin{cases} \varepsilon \cdot [\![\nabla u_h \, n_E]\!]_E & \text{if } E \subset \Omega \setminus \Gamma \\ g_h - \varepsilon \cdot \partial u_h / \partial n & \text{if } E \subset \Gamma_N \\ 0 & \text{if } E \subset \Gamma_D \end{cases} \tag{3.7}$$

Recall that $n_E \perp E$ is one of the two unitary normal vectors whereas $n \perp E \subset \Gamma_N$ denotes the outer unitary normal vector, cf. Chapter 2.

The face residual is also known as gradient jump *or* jump residual. *Note that the element residual r_T is clearly related to the strong form of the differential equation.*

The residuals are often accompanied by a factor which is specific for the PDE and the discretization.

Definition 3.3.2 (Residual scaling factor) *For an element T, define*

$$\alpha_T := \min\{1, \varepsilon^{-1/2} \, h_{min,T}\} \qquad . \tag{3.8}$$

Next the error estimator is defined, and the main result is presented.

Definition 3.3.3 (Residual error estimator) *Define the* local residual error estimator $\eta_{\varepsilon,T}$ *for a tetrahedron T by*

$$\eta_{\varepsilon,\mathrm{R},T} := \left(\alpha_T^2 \cdot \|r_T\|_T^2 + \varepsilon^{-1/2} \cdot \alpha_T \cdot \sum_{E \subset \partial T \setminus \Gamma_D} \|r_E\|_E^2 \right)^{1/2}. \tag{3.9}$$

To shorten the notation, define the local approximation term

$$\zeta_{\varepsilon,T} := \left(\alpha_T^2 \cdot \sum_{T' \subset \omega_T} \|R_{T'} - r_{T'}\|_{T'}^2 + \varepsilon^{-1/2} \cdot \alpha_T \cdot \sum_{E \subset \partial T \cap \Gamma_N} \|g - g_h\|_E^2 \right)^{1/2}$$

that can also be viewed as a consistency error expression. Finally, define the global *terms*

$$\eta_{\varepsilon,\mathrm{R}}^2 := \sum_{T \in \mathcal{T}_h} \eta_{\varepsilon,\mathrm{R},T}^2 \qquad and \qquad \zeta_\varepsilon^2 := \sum_{T \in \mathcal{T}_h} \zeta_{\varepsilon,T}^2 \qquad .$$

Theorem 3.3.4 (Residual error estimation) *The error is bounded locally from below for all $T \in \mathcal{T}_h$ by*

$$\eta_{\varepsilon,\mathrm{R},T} \lesssim \|\|u - u_h\|\|_{\omega_T} + \zeta_{\varepsilon,T} \quad . \tag{3.10}$$

The error is bounded globally from above by

$$\|\|u - u_h\|\| \lesssim m_1(u - u_h, \mathcal{T}_h) \cdot \left[\eta_{\varepsilon,\mathrm{R}}^2 + \zeta_\varepsilon^2\right]^{1/2} \quad . \tag{3.11}$$

Both error bounds are uniform in ε.

Key ideas of the proof: The proof is essentially given in [Kun01e]. Neumann boundary conditions that are not included in that paper can be treated in a straight-forward fashion.

The general methodology to derive the *upper error bound* is similar to that of the Poisson problem, cf. [Kun00]. One basically utilizes the Galerkin orthogonality for a Clément interpolant of the error function, the Cauchy Schwarz inequality and interpolation error estimates. The interpolation estimates have been derived specifically for the anisotropic discretization and the singularly perturbed PDE (3.1). The interpolation estimates also introduce the alignment measure m_1 to (3.11).

The *lower error bound* is proven using the original bubble functions as well as the squeezed bubble functions. The proof is complemented by the corresponding inverse inequalities. All ingredients are designed to properly reflect the singular character of the PDE (3.1).   ■

Remark 3.3.5 Combining the lower and upper error bound yields

$$\eta_{\varepsilon,\mathrm{R}}^2 - c \cdot \zeta_\varepsilon^2 \lesssim \|\|u - u_h\|\| \lesssim m_1^2(u - u_h, \mathcal{T}_h) \cdot \left[\eta_{\varepsilon,\mathrm{R}}^2 + \zeta_\varepsilon^2\right] \quad .$$

Assuming that the approximation term ζ_ε is negligible, one obtains sharp error bounds if the alignment measure $m_1(u - u_h, \mathcal{T}_h)$ is small, which in turn implies that the anisotropic mesh is well suited to the anisotropic solution. For more details we refer to Section 2.3.

Note that in practical applications $m_1(u - u_h, \mathcal{T}_h)$ can be approximated, e.g. by means of a recovered gradient [Kun99, Kun00], see also Section 2.3.   □

Remark 3.3.6 In the original paper [Kun01e], the exact residual R_T is replaced by a piecewise *linear* approximation, in contrast to the *constant* approximation presented here. This results only in minimal differences of the analysis. The version presented above, however, allows to prove an equivalence of the residual error estimator $\eta_{\varepsilon,\mathrm{R},T}$ with the local problem error estimator $\eta_{\varepsilon,\mathrm{D},T}$ presented in the next Section 3.4.   □

3.4   Local problem error estimator

The material of this section originates from [Kun01f].

3.4.1   Motivation and definition of the estimator

The main ideas behind local problem error estimation have been known for a long time [AO00, BR78, BW85, Ver94, Ver96]. Basically the problem is solved locally but with higher accuracy, and the difference between the new solution and the original finite element solution serves as error estimator.

On isotropic meshes there is a large variety of estimators which is underpinned by the literature cited above. For anisotropic discretizations the situation is different. Some of the isotropic estimators also work here, but others fail. Both situations are documented in [Kun01a] for the Poisson equation.

Let us turn now to the singularly perturbed reaction diffusion problem of this chapter. One main question concerns a suitable local (finite dimensional) ansatz space V_T in which to solve the problem. The ansatz spaces used for the Poisson problem are not appropriate. Instead, the ansatz space has to be modified in order to take the singularly perturbed character of the PDE into account. In particular the face bubble functions (contained in the ansatz space V_T) have to be adapted. For isotropic meshes this adaption has been introduced in [Ver98b, Section 3]. The concept has been extended and slightly improved in [Kun01e, Kun01f] to cover anisotropic discretizations as well. There we have coined the term *squeezed face bubble functions*, see also Section 2.2.

In the remainder of this section we define the local ansatz space V_T, the local problem and eventually the error estimator. Section 3.4.2 presents the main results and explains the key ideas and the philosophy behind the proofs. In Section 3.4.3 we first remark on a face-based estimator. Then we address implementational aspects of the error estimator. They do not primarily influence the analysis of the estimator but have major impact on the computational effort and hence on the overall performance of an adaptive algorithm.

All the exposition closely follows [Kun01f].

When designing the error estimator, we first have to decide on the local *subdomain* on which the local problem is to be solved. Here the subdomain is chosen to be the patch ω_T, i.e. the element T and all its face neighbours.

Next the local *ansatz space* V_T has to be specified. To motivate our choice, introduce first the infinite dimensional subspace

$$H_0^1(\omega_T) := \left\{ v \in H^1(\Omega) : \operatorname{supp} v \subseteq \omega_T, \quad v = 0 \text{ on } \partial\omega_T \setminus \Gamma_N \right\} \quad .$$

For an arbitrary function $v \in H_0^1(\omega_T)$ the error then satisfies

$$a(u - u_h, v)\Big|_{\omega_T} = \int_{\omega_T} f \cdot v + \int_{\partial\omega_T \cap \Gamma_N} g \cdot v - \int_{\omega_T} \varepsilon(\nabla u_h)^\top \nabla v - \int_{\omega_T} u_h v \quad . \tag{3.12}$$

The local problem is obtained by approximating the space $H_0^1(\omega_T)$ by some local, finite dimensional space $V_T \subset H_0^1(\omega_T)$. The demands when designing V_T are partly contradicting:

- The local problem should be cheap to solve which implies a small local space V_T.
- The space V_T should be rich enough to extract information on the error $e := u - u_h$.
- From an analytical point of view, we want to obtain certain equivalences between the residual error estimator $\eta_{\varepsilon,R,T}$ and the local problem estimator $\eta_{\varepsilon,D,T}$ defined below. This implies restrictions on the dimension and the functions of the local space V_T.

The local space that we use here is spanned by certain bubble functions, cf. Section 2.2. One basis function is formed by the element bubble function b_T. Furthermore one needs face bubble functions. However, the standard functions b_E are unsuitable because of the singularly perturbed character of the PDE. Hence we have to employ the *squeezed* face bubble functions b_{E,γ_E}. The subsequent analysis suggests to set the squeezing parameter to

$$\gamma_E := \min\left\{ 1, \frac{h_{min,E}}{h_E}, \frac{\sqrt{\varepsilon}}{h_E} \right\} \quad . \tag{3.13}$$

The local space V_T is now defined by

$$V_T := \text{span}\{b_T, b_{E,\gamma_E} : E \subset \partial T \setminus \Gamma_D\} \qquad . \qquad (3.14)$$

For interior elements this space has dimension 5.

The local problem and the error estimator can be formulated most conveniently by means of the (approximate) residuals.

Definition 3.4.1 (Local Dirichlet problem error estimator)
Find a solution $e_T \in V_T$ of the local variational problem:

$$a(e_T, v_T) \equiv \int_{\omega_T} \varepsilon(\nabla e_T)^\top \nabla v_T + e_T\, v_T$$

$$= \sum_{T' \subset \omega_T} \int_{T'} r_{T'} \cdot v_T + \sum_{E \subset \partial T \setminus \Gamma_D} \int_E r_E \cdot v_T \qquad \forall\, v_T \in V_T. \qquad (3.15)$$

The local *and* global error estimators *then become*

$$\eta_{\varepsilon,D,T} := |\!|\!| e_T |\!|\!|_{\omega_T} \qquad and \qquad \eta_{\varepsilon,D}^2 := \sum_{T \in \mathcal{T}_h} \eta_{\varepsilon,D,T}^2 \qquad . \qquad (3.16)$$

Note that the particular choice of the local ansatz space V_T (namely $v_T = 0$ on $\partial \omega_T \setminus \partial T$) reduces certain boundary integrals and norms. An equivalent, alternative formulation of the local problem is derived by partial integration: Find $e_T \in V_T$ such that

$$a(e_T, v_T) = a(u - u_h, v_T) - \sum_{T' \subset \omega_T} \int_{T'} (R_{T'} - r_{T'})\, v_T - \int_{\partial T \cap \Gamma_N} (g - g_h)\, v_T \qquad \forall\, v_T \in V_T.$$

3.4.2   Main results

The methodology of the error estimator partly utilizes ideas that have already been introduced for the anisotropic local problem estimator for the Poisson problem [Kun01a], and for the anisotropic residual estimator for a singularly perturbed reaction diffusion equation [Kun01e]. The actual ingredients differ of course.

To start with, Lemma 3.4.2 is the cornerstone for the error analysis. It gives two relations that are similar to inverse inequalities. With its help one can establish equivalences between the local problem error estimator $\eta_{\varepsilon,D,T}$ and the residual error estimator $\eta_{\varepsilon,R,T}$, see Theorem 3.4.3 below. The actual error bounds of Theorem 3.4.4 can then be concluded easily. Finally, for the practical computation of the local problem error estimator one wants to employ a basis of the local space V_T that is *robust* with respect to the perturbation parameter ε. In other words, the condition number of the local stiffness matrix corresponding to (3.15) does not deteriorate for small values of ε. Such a stable basis is presented, and the stability is given by Theorem 3.4.5.

Lemma 3.4.2 *The following relations hold for all $v_T \in V_T$.*

$$\|v_T\|_{\omega_T} \lesssim h_{min,T} \cdot \|\nabla v_T\|_{\omega_T} \qquad (3.17)$$

$$\|v_T\|_E \lesssim h_E^{-1/2}\, \gamma_E^{-1/2} \cdot \min\{h_{min,T}, \gamma_E\, h_E\} \cdot \|\nabla v_T\|_{\omega_T} \qquad \forall\, E \subset \partial T \qquad . \qquad (3.18)$$

The inequalities are uniform in the squeezing parameters $\gamma_E \in (0, 1]$ which define the space V_T.

If T has at least two Neumann boundary faces then the constants in (3.17), (3.18) can depend on the shape of the Neumann boundary (but do not depend on the triangulation $\mathcal{T}_h$ nor on T). More precisely, this Neumann boundary forms an edge at T, and the angle between the Neumann faces at this edge determine the constants. The smaller this angle, the worse the constants may be.

Key ideas of the proof: The technical proof is given in [Kun01f, Lemma 4.2].

Both relations are similar to inverse inequalities. The particular technicalities arise from the great variability of the local space V_T (i.e. V_T depends on the shape of all elements of the patch ω_T, and on the squeezing parameters γ_E). ∎

Next we show certain equivalences between the local problem error estimator and the residual error estimator of the previous section.

Theorem 3.4.3 (Equivalence with residual error estimator) *The local problem error estimator $\eta_{\varepsilon,\mathrm{D},T}$ is equivalent to the residual error estimator $\eta_{\varepsilon,\mathrm{R},T}$ in the following sense:*

$$\eta_{\varepsilon,\mathrm{D},T}^2 \;\lesssim\; \sum_{T' \subset \omega_T} \eta_{\varepsilon,\mathrm{R},T'}^2 \tag{3.19}$$

$$\eta_{\varepsilon,\mathrm{R},T}^2 \;\lesssim\; \sum_{T' \subset \omega_T} \eta_{\varepsilon,\mathrm{D},T'}^2 \tag{3.20}$$

$$\eta_{\varepsilon,\mathrm{R}} \;\sim\; \eta_{\varepsilon,\mathrm{D}} \quad . \tag{3.21}$$

All inequalities are uniform in ε.

If T has at least two Neumann boundary faces then the constants in (3.19) and (3.21) can depend on the shape of the Neumann boundary (but do not depend on the triangulation $\mathcal{T}_h$ nor on T).

Key ideas of the proof: Inequality (3.19) hinges on the previous Lemma 3.4.2. The proof of (3.20) employs bubble functions and inverse inequalities. Details are given in [Kun01f, Theorem 4.3].

Roughly speaking, the local space V_T has to correspond to the data that define the residual error estimator $\eta_{\varepsilon,\mathrm{R},T}$. The element bubble function $b_T \in V_T$ captures contributions from the element residual r_T while the squeezed face bubble functions $b_{E,\gamma_E} \in V_T$ capture contributions from the face residuals r_E. ∎

The actual error bounds now follow easily.

Theorem 3.4.4 (Local problem error estimation)
The error is bounded locally from below by

$$\eta_{\varepsilon,\mathrm{D},T} \;\leq\; \left\|\!\left\|\!\left\| u - u_h \right\|\!\right\|\!\right\|_{\omega_T} + c \cdot \zeta_{\varepsilon,T} \qquad \forall\, T \in \mathcal{T}_h \quad . \tag{3.22}$$

The error is bounded globally from above by

$$\left\|\!\left\|\!\left\| u - u_h \right\|\!\right\|\!\right\| \;\lesssim\; m_1(u - u_h, \mathcal{T}_h) \cdot \left[\eta_{\varepsilon,\mathrm{D}}^2 + \zeta_\varepsilon^2 \right]^{1/2} \quad . \tag{3.23}$$

Both inequalities are uniform in ε.

The lower error bound (3.22) is a strict inequality where the only constant c is at the data approximation term $\zeta_{\varepsilon,T}$. As always, this constant c is independent of ε, T, u and u_h. However, if T has at least two Neumann boundary faces then c can depend on the shape of the Neumann boundary (but does not depend on the triangulation $\mathcal{T}_h$ nor on T).

Key ideas of the proof: The proof is given in [Kun01f].

The lower error bound (3.22) is a simple consequence of the approximation of the subproblem (3.12) by the finite dimensional space V_T.

The upper error bound rests on the equivalence (3.20) of the local problem error estimator with the residual error estimator $\eta_{\varepsilon,\mathrm{R},T}$, and the error bound (3.11) for the latter estimator. $\blacksquare$

When actually computing the local problem error estimator, one has to solve the finite dimensional problem (3.4.1). With a fixed basis of the local space V_T, this leads to a system of linear equations with a local FEM stiffness matrix K_T.

The question now is how to choose the basis of V_T. An obvious choice would consist of that functions that define V_T in (3.14), i.e. consisting of the element bubble function b_T and (generically four) squeezed face bubble functions b_{E,γ_E}. Then, however, the condition number of the local stiffness matrix K_T grows as the perturbation parameter ε tends to zero. In other words, the aforementioned basis of V_T is unstable.

Therefore we present here a basis Φ of V_T which guarantees a stable solution of the local problem (3.15). Recall that the local ansatz space is $V_T = \mathrm{span}\{b_T, b_{E,\gamma_E} : E \subset \partial T \setminus \Gamma_\mathrm{D}\}$. As a basis of V_T we choose

$$\Phi := \left(b_T \,,\, \gamma_E^{-1/2} \cdot b_{E,\gamma} \,:\, E \subset \partial T \setminus \Gamma_\mathrm{D} \right) \qquad . \tag{3.24}$$

Via the standard finite element isomorphism this local basis defines a corresponding local stiffness matrix K_T.

Theorem 3.4.5 (Stable basis) *The basis (3.24) of V_T is stable, i.e. the condition number $\kappa(K_T)$ of the local problem stiffness matrix K_T is bounded uniformly in ε and T:*

$$\kappa(K_T) \sim 1 \qquad \forall\, T \in \mathcal{T}_h \qquad .$$

Proof: See [Kun01f, Theorem 4.7]. $\blacksquare$

3.4.3 Remarks

A further, face based local problem error estimator

With the methodology presented so far one can derive further local problem error estimators. This will be demonstrated here for a *face based* local problem error estimator. Such an estimator can be advantageous when other ingredients of an adaptive algorithm are face based too (e.g. the refinement procedure).

The local space associated to a face E is set to

$$V_E := \mathrm{span}\{b_{E,\gamma_E} \text{ if } E \not\subset \Gamma_\mathrm{D} \,,\, b_T \,\forall\, T \subset \omega_E\} \qquad ,$$

i.e. V_E is three dimensional for interior faces E. The local problem is: Find $e_E \in V_E$ such that

$$a(e_E, v_E) = \sum_{T \subset \omega_E} \int_T r_T \cdot v_E + \int_E r_E \cdot v_E \qquad \forall\, v_E \in V_E \qquad .$$

The *local face error estimator* and the *face based approximation term* then become

$$\eta_{\varepsilon,\mathrm{D},E} := \|\!\|\!| e_E \|\!\|\!|_{\omega_E}$$

$$\zeta_{\varepsilon,E} := \alpha_E \left(\sum_{T \subset \omega_E} \|R_T - r_T\|_T^2 \right)^{1/2} + \varepsilon^{-1/4} \alpha_E^{1/2} \cdot \|g - g_h\|_{E \cap \Gamma_\mathrm{N}} \qquad ,$$

where $\alpha_E := (\alpha_{T_1} + \alpha_{T_2})/2$ is the mean of the two element terms α_{T_i}, with $E = T_1 \cap T_2$.

Again an alternative, equivalent description of the local problem is possible and advantageous: Find $e_E \in V_E$ such that

$$a(e_E, v_E) \;=\; a(u - u_h, v_E) - \sum_{T \subset \omega_E} \int_T (R_T - r_T)\, v_E \; - \int_{E \cap \Gamma_N} (g - g_h)\, v_E \qquad \forall\, v_E \in V_E.$$

Then the following error bounds can be obtained.

Theorem 3.4.6 (Face based local problem error estimator)
The error is bounded locally from below for all faces E of $\mathcal{T}_h$ by

$$\eta_{\varepsilon, D, E} \;\leq\; \vertiii{u - u_h}_{\omega_E} + c \cdot \zeta_{\varepsilon, E} \qquad .$$

The error is bounded globally from above by

$$\vertiii{u - u_h} \;\lesssim\; m_1(u - u_h, \mathcal{T}_h) \cdot \left(\sum_{E \in \mathcal{T}_h} \eta_{\varepsilon, D, E}^2 + \zeta_{\varepsilon, E}^2 \right)^{1/2} \qquad .$$

All relations are uniform in ε.

Implementational aspects and complexity

It is a major demand that the local problem can be constructed and solved as fast as possible since usually the error estimation is as expensive as the assembly of the global finite element stiffness matrix and the solution process for u_h. Therefore one encounters two main problems when applying our error estimator. Both difficulties are related to the computation of the local stiffness matrix K_T which arises from the bilinear form $a(\cdot, \cdot)$. Recall first the basis Φ of V_T as defined by (3.24).

Problem 1: The support of the squeezed face bubble function $b_{E,\gamma}$ is not ω_E but some (γ–dependent) part of it. For example the computation of $a(b_{E,\gamma}, \cdot)$ implies a comparatively complicated domain of integration. In particular when evaluating $a(b_{E_1,\gamma_{E_1}}, b_{E_2,\gamma_{E_2}})$ for two different squeezed face bubble functions, the domain of integration becomes

$$\operatorname{supp}(b_{E_1,\gamma_{E_1}}) \cap \operatorname{supp}(b_{E_2,\gamma_{E_2}})$$

which might be empty, or a single tetrahedron, or the union of two tetrahedra, depending on γ_{E_1} and γ_{E_2} (cf. also Figures 2.2.2 and 2.2.3). Even to determine and describe the domain of integration is not trivial, save the actual integration.

Remedy: We modify the parameter for the squeezed face bubble function to be

$$\tilde{\gamma}_E := \min\left\{ \frac{1}{4}, \frac{h_{min,E}}{h_E}, \frac{\sqrt{\varepsilon}}{h_E} \right\} \equiv \min\left\{ \frac{1}{4}, \gamma_E \right\} \sim \gamma_E \qquad .$$

Then all results remain valid, only the inequality constants may be slightly worse (but they are still uniform in ε). The main advantage now is that

$$\operatorname{supp}(b_{E_1,\tilde{\gamma}_{E_1}}) \cap \operatorname{supp}(b_{E_2,\tilde{\gamma}_{E_2}}) = \emptyset \qquad .$$

Hence the computation of the modified local matrix $\tilde{K}_T$ is less expensive, as the matrix now contains several zero entries. Even more, the sparsity pattern

$$\tilde{K}_T = \tilde{K}_T^\top = \begin{bmatrix} * & * & * & * & * \\ * & * & 0 & 0 & 0 \\ * & 0 & * & 0 & 0 \\ * & 0 & 0 & * & 0 \\ * & 0 & 0 & 0 & * \end{bmatrix}$$

allows a particularly fast and simple solution of the local problem.

Problem 2: The basis functions of V_T are polynomials of a relatively high degree. Hence numerical integration rules to compute $a(\cdot, \cdot)$ are far too expensive and thus unsuitable.

Remedy: We propose a direct computation of the desired values. This procedure involves geometry data of the element T and the subdomain ω_T as well as integrals that are precomputed analytically. The effort is thus reduced to less than 10% of the effort required for numerical integration.

Furthermore we note that our procedure for the direct computation is very similar to the computation of the local problem for the *Poisson equation*, cf. [Kun01a]. The computational effort is roughly the same, i.e. the singularly perturbed character of the PDE here is no disadvantage.

Details are given in [Kun01f, Section 5].

3.5 H^1 seminorm error estimators

The material of this section originates from [Kun01d]. Unfortunately this preprint contains a (minor) mistake concerning the approximation of the element residual. This has been corrected in [Kun01c] which has been submitted to the IMA Journal of Numerical Analysis (currently under review).

In this section we do not consider the energy norm which is closely related to the corresponding bilinear form. Here we aim at error bounds in the H^1 *seminorm*. Partly this work has been motivated by [KS01].

From definition (3.4) of the energy norm $\|\|\cdot\|\|$ one infers a trivial relation with the H^1 seminorm $\|\nabla \cdot \|$, namely

$$\|\nabla v\| \leq \varepsilon^{-1/2} \|\|v\|\| \leq \|\nabla v\| + \varepsilon^{-1/2}\|v\|.$$

Hence we observe an obvious possibility to define a first crude error estimator for the H^1 seminorm: Take an energy norm error estimator and scale it by $\varepsilon^{-1/2}$. Then the corresponding upper error bound (for the H^1 seminorm) holds immediately. Only the lower error bound contains an additional L^2 term.

Naturally we expected better results from a detailed analysis. Several variations of error estimators have been investigated:

- two residual error estimators (with different scaling of the residuals),
- three local problem error estimators that differ either by the local problem or by the norm that defines the estimator.

Much to our surprise we failed to improve the first crude error bounds described above. Even more, the choice of a suitable local problem has been far from obvious.

Because of these somewhat unsatisfactory results we present only the (apparently) best residual error estimator and the best local problem error estimator. Roughly speaking, they are obtained from the corresponding energy norm error estimators by a simple scaling with $\varepsilon^{-1/2}$. Hence the philosophy and the remarks are similar to the ones of the previous two sections. Therefore we also pare down our exposition to the minimum.

3.5.1 A residual error estimator

Define the exact residual R_T and the approximate residuals r_T, r_E as before in Definition 3.3.1 on page 25. The residual scaling factor α_T is the same as before in Definition 3.3.2. Then we are ready to define the error estimator.

Definition 3.5.1 (H^1 residual error estimator)
For a tetrahedron T, define the residual error estimator *by*

$$\eta_{H^1,\mathrm{R},T} := \left(\varepsilon^{-1}\alpha_T^2 \cdot \|r_T\|_T^2 + \varepsilon^{-3/2}\alpha_T \cdot \sum_{E \subset \partial T \setminus \Gamma_\mathrm{D}} \|r_E\|_E^2 \right)^{1/2}. \tag{3.25}$$

To shorten the notation, introduce the local residual approximation term

$$\zeta_{H^1,T} := \left(\varepsilon^{-1}\alpha_T^2 \cdot \sum_{T' \subset \omega_T} \|R_{T'} - r_{T'}\|_{T'}^2 + \varepsilon^{-3/2}\alpha_T \cdot \sum_{E \subset \partial T \cap \Gamma_\mathrm{N}} \|g - g_h\|_E^2 \right)^{1/2}. \tag{3.26}$$

Finally, define the global *terms*

$$\eta_{H^1,\mathrm{R}}^2 := \sum_{T \in \mathcal{T}_h} \eta_{H^1,\mathrm{R},T}^2 \qquad and \qquad \zeta_{H^1}^2 := \sum_{T \in \mathcal{T}_h} \zeta_{H^1,T}^2 \quad .$$

Note that all terms are the $\varepsilon^{-1/2}$ scaled variations of the corresponding items for the energy norm, cf. Section 3.3. The error bounds read as follows.

Theorem 3.5.2 (H^1 residual error estimation)
The error is bounded locally from below for all $T \in \mathcal{T}_h$ by

$$\eta_{H^1,\mathrm{R},T} \lesssim \|\nabla(u - u_h)\|_{\omega_T} + \varepsilon^{-1/2}\alpha_T \cdot \|u - u_h\|_{\omega_T} + \zeta_{H^1,T} \quad . \tag{3.27}$$

The error is bounded globally from above by

$$\|\nabla(u - u_h)\|_\Omega \lesssim m_1(u - u_h, \mathcal{T}_h) \cdot \left[\eta_{H^1,\mathrm{R}}^2 + \zeta_{H^1}^2 \right]^{1/2} \quad . \tag{3.28}$$

Both error bounds are uniform in ε.

Proof: See [Kun01c]. ∎

When comparing the lower error bound (3.27) with the corresponding error bounds for the energy norm (cf. Theorems 3.3.4 and 3.4.4), we observe the additional term $\varepsilon^{-1/2}\alpha_T \cdot \|u - u_h\|_{\omega_T}$. As already mentioned above, it stems solely from the difference between the energy norm and the H^1 seminorm.

3.5.2 A local problem error estimator

The basic settings are very similar to the ones of Section 3.4. The local problem is posed on the subdomain ω_T. The local space is given exactly as in (3.14) by

$$V_T := \mathrm{span}\{b_T, b_{E,\gamma_E} : E \subset \partial T \setminus \Gamma_\mathrm{D}\},$$

with the squeezing parameter γ_E being

$$\gamma_E := \min\{1, \varepsilon^{1/2}/h_E, h_{min,E}/h_E\}$$

as before. The local problem and the error estimator are then defined as follows.

Definition 3.5.3 (H^1 local problem error estimator)
Find a solution $e_T \in V_T$ of the local variational problem:

$$a(e_T, v_T) = \sum_{T' \subset \omega_T} \int_{T'} r_{T'} \cdot v_T + \sum_{E \subset \partial T \backslash \Gamma_D} \int_E r_E \cdot v_T \qquad \forall\, v_T \in V_T. \tag{3.29}$$

The local *and* global error estimators *then* become

$$\eta_{H^1,\mathrm{D},T} := \varepsilon^{-1/2} \cdot \|\|e_T\|\|_{\omega_T} \qquad and \qquad \eta_{H^1,\mathrm{D}}^2 := \sum_{T \in \mathcal{T}_h} \eta_{H^1,\mathrm{D},T}^2 \quad . \tag{3.30}$$

Both the local problem and the error estimator are again the $\varepsilon^{-1/2}$ scaled variations of the corresponding items for the energy norm, cf. Section 3.4. Hence we obtain a similar equivalence between the residual error estimator and the local problem error estimator as in Theorem 3.4.3.

Theorem 3.5.4 (Equivalence with residual error estimator)   *The local problem error estimator $\eta_{H^1,\mathrm{D},T}$ is equivalent to the residual error estimator $\eta_{H^1,\mathrm{R},T}$ in the following sense:*

$$\eta_{H^1,\mathrm{D},T}^2 \lesssim \sum_{T' \subset \omega_T} \eta_{H^1,\mathrm{R},T'}^2 \tag{3.31}$$

$$\eta_{H^1,\mathrm{R},T}^2 \lesssim \sum_{T' \subset \omega_T} \eta_{H^1,\mathrm{D},T'}^2 \tag{3.32}$$

$$\eta_{H^1,\mathrm{R}} \sim \eta_{H^1,\mathrm{D}} \quad . \tag{3.33}$$

All inequalities are uniform in ε.
   If T has at least two Neumann boundary faces then the constants in (3.31) and (3.33) can depend on the shape of the Neumann boundary (but does not depend on the triangulation $\mathcal{T}_h$ nor on T).

The actual error bounds follow easily.

Theorem 3.5.5 (H^1 local problem error estimation)
The error is bounded locally from below for all $T \in \mathcal{T}_h$ by

$$\eta_{H^1,\mathrm{D},T} \le \|\nabla(u - u_h)\|_{\omega_T} + \varepsilon^{-1/2} \cdot \|u - u_h\|_{\omega_T} + c\,\zeta_{H^1,T} \quad . \tag{3.34}$$

The error is bounded globally from above by

$$\|\nabla(u - u_h)\|_\Omega \lesssim m_1(u - u_h, \mathcal{T}_h) \cdot \left[\eta_{H^1,\mathrm{D}}^2 + \zeta_{H^1}^2\right]^{1/2} \quad . \tag{3.35}$$

Both inequalities are uniform in ε.
   The lower error bound (3.34) is a strict inequality where the only constant c is at the residual approximation term $\zeta_{H^1,T}$. As always, this constant c is independent of ε, T, u and u_h. However, if T has at least two Neumann boundary faces then c can depend on the shape of the Neumann boundary (but does not depend on the triangulation $\mathcal{T}_h$ nor on T).

3.5.3 Remarks on different estimators

By using different scaling factors we obtained a second, modified residual error estimator, cf. [Kun01c, Section 4.2]. The resulting error bounds are similar *in structure* but with different factors. In numerical experiments the modified residual error estimator *overestimated* the global error considerably; hence the estimator is unsuitable.

In a similar fashion we investigated alternative local problem error estimators. Two modifications have been proposed, cf. [Kun01c, Section 5.2].

1. Solve the same local problem (3.29). Since we aim at the H^1 seminorm, it seems natural to to take this H^1 seminorm of the local solution. Hence define the first modified local problem error estimator by

$$\tilde{\eta}_{H^1,\mathrm{D},T} := \|\nabla e_T\|_{\omega_T}.$$

2. Modify the local problem by considering only the Poisson part of the PDE (3.2): Find $\breve{e}_T \in V_T$ of the local variational problem

$$\varepsilon \int_{\omega_T} \nabla \breve{e}_T \nabla v_T \;=\; \sum_{T' \subset \omega_T} \int_{T'} r_{T'} \cdot v_T \;+\; \sum_{E \subset \partial T \backslash \Gamma_\mathrm{D}} \int_E r_E \cdot v_T \qquad \forall\, v_T \in V_T \quad .$$

Define the second modified local problem error estimator by

$$\breve{\eta}_{H^1,\mathrm{D},T} := \|\nabla \breve{e}_T\|_{\omega_T}.$$

From a theoretical point of view, the upper and/or lower error bounds are less fortunate. The numerical experiments confirm that the second modified estimator is unsuitable for practical applications. For the first modification more research is necessary.

3.6 Numerical experiments

The numerical experiments of this section are contained in [Kun01e, Kun01f, Kun01c]. The underlying problems are the same for each of the estimators. Hence we present the numerical experiment for all estimators together.

Let us consider the 3D model problem

$$\begin{aligned} -\varepsilon \Delta u + u &= 0 && \text{in } \Omega := (0,1)^3 \\ u &= u_0 && \text{on } \Gamma_\mathrm{D} := \partial\Omega. \end{aligned}$$

The Dirichlet boundary value u_0 is chosen such that the exact analytical solution becomes

$$u = e^{-x/\sqrt{\varepsilon}} + e^{-y/\sqrt{\varepsilon}} + e^{-z/\sqrt{\varepsilon}} \quad .$$

This prescribed solution features three typical exponential boundary layers along the planes $x = 0$, $y = 0$, and $z = 0$. The boundary layer width is $\mathcal{O}(\sqrt{\varepsilon}|\ln\sqrt{\varepsilon}|)$. The value of the perturbation parameter $\varepsilon \ll 1$ will be specified later.

We apply the finite element method with a sequence of meshes $\mathcal{T}_k$, each of which is the tensor product of three one–dimensional Bakhvalov–like meshes [Bak69] with 2^k intervals in $[0,1]$, $k = 1 \ldots 6$. To describe the 1D nodal distribution properly, denote the transition point of the boundary layer by $\tau := \sqrt{\varepsilon}|\ln\sqrt{\varepsilon}|$ which is related to the layer width. Then 2^{k-1} nodes are *exponentially* distributed in the boundary layer interval $[0,\tau]$ whereas the remaining interval $[\tau, 1]$ is divided into 2^{k-1} *equidistant* intervals, cf. Figure 3.6.3. More precisely, the (1D) nodal coordinate of the m-th node is

$$
x_m := \begin{cases} -\beta\sqrt{\varepsilon}\ln\left[1 - \dfrac{m}{2^{k-1}}(1 - e^{-\tau/\beta/\sqrt{\varepsilon}})\right] & \text{for } m = 0\ldots 2^{k-1}, \beta = 3/2 \\[2ex] \tau + (1 - \tau)\cdot\left(\dfrac{m}{2^{k-1}} - 1\right) & \text{for } m = 2^{k-1}+1\ldots 2^k \end{cases}
$$

Note that the original (1D) Bakhvalov mesh utilizes a slightly different transition point τ. Furthermore we do not know whether these tensor product type meshes are optimal (which, of course, also depends on the optimality criterion).

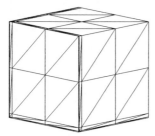 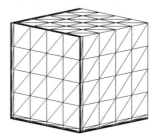

Figure 3.6.3: Mesh $\mathcal{T}_2$ – Mesh $\mathcal{T}_3$

The exposition of the results is twofold.

- For the two *energy norm* error estimators $\eta_{\varepsilon,\mathrm{R},T}$ and $\eta_{\varepsilon,\mathrm{D},T}$ we use the value $\varepsilon = 10^{-4}$ as in [Kun01e, Kun01f]. The results for $\varepsilon = 10^{-8}$ are very similar but they are not documented here.

- For the two H^1 *seminorm* error estimators $\eta_{H^1,\mathrm{R},T}$ and $\eta_{H^1,\mathrm{D},T}$ we employ the value $\varepsilon = 10^{-8}$ as in [Kun01c]. We have chosen this (smaller) value because the (negative) results for the modified estimators are more convincing there.

3.6.1 Energy norm error estimates

The results for the local problem error estimator are taken from [Kun01f]. The residual error estimator here is exactly as defined by (3.9), i.e. using a *constant* approximation of the exact element residual R_T. This has the advantage that we can compare both estimators. The slight disadvantage is that in [Kun01e] the residual error estimator has been investigated numerically using a *linear* approximation of R_T.

Recall that the perturbation parameter is $\varepsilon = 10^{-4}$ here. We start with some useful information about the meshes and the solution. The table below gives the number of elements, the maximum aspect ratio $\max_{T\in\mathcal{T}_h} h_{1,T}/h_{3,T}$ of the mesh, and the value of the alignment measure.

Mesh	# Elements	Aspect ratio	$m_1(u - u_h, \mathcal{T}_k)$
$\mathcal{T}_1$	48	29.4	1.55
$\mathcal{T}_2$	384	69.5	1.62
$\mathcal{T}_3$	3 072	82.6	1.69
$\mathcal{T}_4$	24 576	88.6	1.88
$\mathcal{T}_5$	196 608	91.5	2.37
$\mathcal{T}_6$	1 572 864	92.9	3.04

Since the size of m_1 is comparatively small and grows only mildly, the chosen meshes discretize the problem sufficiently well.

Decrease of the error and the error estimators

The following table lists the error as well as both error estimators. The last column contains the approximation term ζ_ε.

Mesh	$\|\|\|u - u_h\|\|\|$	$\eta_{\varepsilon,\mathrm{R}}$	$\eta_{\varepsilon,\mathrm{D}}$	ζ_ε
$\mathcal{T}_1$	$0.154E + 0$	$0.215E + 0$	$0.102E + 0$	$0.163E + 0$
$\mathcal{T}_2$	$0.536E - 1$	$0.197E + 0$	$0.407E - 1$	$0.721E - 1$
$\mathcal{T}_3$	$0.229E - 1$	$0.115E + 0$	$0.186E - 1$	$0.271E - 1$
$\mathcal{T}_4$	$0.110E - 1$	$0.612E - 1$	$0.926E - 2$	$0.963E - 2$
$\mathcal{T}_5$	$0.553E - 2$	$0.316E - 1$	$0.472E - 2$	$0.352E - 2$
$\mathcal{T}_6$	$0.282E - 2$	$0.161E - 1$	$0.239E - 2$	$0.148E - 2$

For a better understanding all values are displayed in Figure 3.6.4. Observe first a convergence rate of the error $\|\|\|u - u_h\|\|\|$ of approximately $N^{-0.324}$, with N being the number of elements. This is almost the optimal rate of $N^{-1/3}$ which indicates that the meshes under consideration discretize the singular problem well. Secondly, both error estimators decrease at a very similar rate. The local problem estimator is much closer to the error, while the residual estimator overestimates the error by a factor of roughly 5. Lastly, the approximation term ζ_ε is comparatively large, in particular on coarse meshes. This is primarily due to the *constant* approximation of the exact element residual R_T.

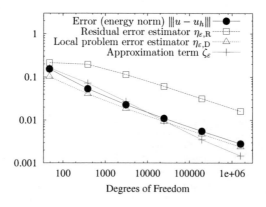

Figure 3.6.4: Energy norm: Error, error estimators, and approximation term.

Upper and lower error bounds

Next we investigate the main results of Theorems 3.3.4 and 3.4.4, namely the upper and lower error bounds. For a compact presentation define the ratios

$$q_{\text{up}} \quad := \quad \frac{\|\|u - u_h\|\|}{m_1(u - u_h, \mathcal{T}_h) \cdot (\eta_{\varepsilon,*}^2 + \zeta_\varepsilon^2)^{1/2}}$$

$$q_{\text{low}} \quad := \quad \max_{T \in \mathcal{T}_k} \frac{\eta_{\varepsilon,*,T}}{\|\|u - u_h\|\|_{\omega_T} + \zeta_{\varepsilon,T}}$$

where $\eta_{\varepsilon,*}$ and $\eta_{\varepsilon,*,T}$ stand either for the residual estimator or the local problem error estimator. Following both aforementioned theorems, these ratios have to be bounded from above, i.e.

$$q_{\text{up}} \lesssim 1 \quad \text{and} \quad q_{\text{low}} \lesssim 1.$$

The first property is often referred to as *reliability* while the latter property implies *efficiency* of the estimator.

The table below gives the corresponding values, and Figure 3.6.5 displays the values graphically. The theoretical predictions are confirmed. Moreover, the absolute sizes of q_{up} and q_{low} are comparable to those for the Poisson equation, cf. [Kun00, Kun01a].

Mesh	Residual error estimator		Local problem error estimator	
	q_{up}	q_{low}	q_{up}	q_{low}
$\mathcal{T}_1$	0.368	0.655	0.517	0.308
$\mathcal{T}_2$	0.157	1.657	0.399	0.321
$\mathcal{T}_3$	0.115	2.842	0.413	0.415
$\mathcal{T}_4$	0.094	3.263	0.437	0.480
$\mathcal{T}_5$	0.073	3.511	0.396	0.507
$\mathcal{T}_6$	0.057	3.575	0.330	0.511

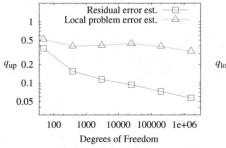 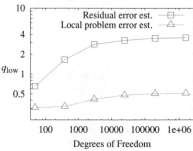

Figure 3.6.5: Left: Upper error bound: $q_{\text{up}} \lesssim 1$
Right: Lower error bound: $q_{\text{low}} \lesssim 1$

Equivalence of the error estimators

In the last table we examine the equivalence of the local problem error estimator and the residual error estimator, as described by Theorem 3.4.3. Define the ratios

$$q_{D/R} := \max_{T \in \mathcal{T}_k} \frac{\eta_{\varepsilon,D,T}}{\left(\sum_{T' \subset \omega_T} \eta_{\varepsilon,R,T'}^2 \right)^{1/2}} \quad \text{and} \quad q_{R/D} := \max_{T \in \mathcal{T}_k} \frac{\eta_{\varepsilon,R,T}}{\left(\sum_{T' \subset \omega_T} \eta_{\varepsilon,D,T'}^2 \right)^{1/2}}.$$

Theorem 3.4.3 implies

$$q_{D/R} \lesssim 1 \quad \text{and} \quad q_{R/D} \lesssim 1.$$

This theoretical prediction is impressively confirmed by the next table which is taken from [Kun01f]. The data are again visualised.

Mesh	$q_{D/R}$	$q_{R/D}$
$\mathcal{T}_1$	0.302	1.556
$\mathcal{T}_2$	0.443	4.278
$\mathcal{T}_3$	0.507	4.956
$\mathcal{T}_4$	0.386	4.851
$\mathcal{T}_5$	0.275	4.708
$\mathcal{T}_6$	0.202	4.770

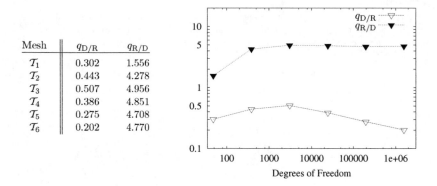

Figure 3.6.6: Energy norm: Equivalence of error estimators: $q_{D/R} \lesssim 1$ and $q_{R/D} \lesssim 1$

Summarising, both estimators behave similarly on a qualitative scale whereas from a quantitative viewpoint one observes roughly $\eta_{\varepsilon,R} \approx 4 \cdot \eta_{\varepsilon,D}$. Furthermore the residual error estimator $\eta_{\varepsilon,R}$ overestimates the true error more than the local problem error estimator $\eta_{\varepsilon,D}$ does. This indeed can be expected since the derivation of $\eta_{\varepsilon,R}$ requires more intermediate steps (such as interpolation estimates and Cauchy Schwarz inequalities).

3.6.2 H^1 seminorm error estimates

The results for the H^1 seminorm error estimators are taken from [Kun01c].

Recall first the value $\varepsilon = 10^{-8}$ of the perturbation parameter. The numerical results and their exposition will be in a similar fashion as above. We start with some information on the discretization. The small values of the alignment measure m_1 indicate that the anisotropic mesh is well suited.

Mesh	# Elements	Aspect ratio	$m_1(u - u_h, \mathcal{T}_k)$
$\mathcal{T}_1$	48	1534	1.49
$\mathcal{T}_2$	384	6815	1.49
$\mathcal{T}_3$	3 072	8206	1.52
$\mathcal{T}_4$	24 576	8838	1.52
$\mathcal{T}_5$	196 608	9142	1.54
$\mathcal{T}_6$	1 572 864	9291	1.60

Decrease of the error and the error estimators

Next we present the H^1 seminorm error, both error estimators, and the approximation term. The graphical illustration of Figure 3.6.7 complements the table. The numerical convergence rate of approximately $N^{-0.34}$ is close to the optimal value which confirms the expediency of the anisotropic meshes $\mathcal{T}_k$. Both error estimators behave similarly on a qualitative scale. The residual error estimator overestimates the error while the local problem error estimator is much closer to the error.

Mesh	$\|\nabla(u - u_h)\|$	$\eta_{H^1,\mathrm{R}}$	$\eta_{H^1,\mathrm{D}}$	ζ_{H^1}
$\mathcal{T}_1$	$9.97E+1$	$2.85E+2$	$1.82E+2$	$2.18E+2$
$\mathcal{T}_2$	$5.51E+1$	$2.24E+2$	$5.75E+1$	$8.36E+1$
$\mathcal{T}_3$	$2.47E+1$	$1.23E+2$	$2.27E+1$	$3.05E+1$
$\mathcal{T}_4$	$1.18E+1$	$6.48E+1$	$1.04E+1$	$1.10E+1$
$\mathcal{T}_5$	$5.78E+0$	$3.34E+1$	$5.04E+0$	$3.91E+0$
$\mathcal{T}_6$	$2.87E+0$	$1.69E+1$	$2.50E+0$	$1.38E+0$

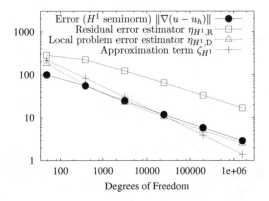

Figure 3.6.7: H^1 seminorm: Error, error estimators, and approximation term.

Upper and lower error bounds

Again we investigate the main results of Theorems 3.5.2 and 3.5.5. Corresponding to the upper and lower error bounds, we define the ratios

$$q_{\mathrm{up}} := \frac{\|\nabla(u - u_h)\|}{m_1(u - u_h, \mathcal{T}_h) \cdot (\eta_{H^1,*}^2 + \zeta_{H^1}^2)^{1/2}}$$

$$q_{\mathrm{low}} := \max_{T \in \mathcal{T}_k} \begin{cases} \dfrac{\eta_{H^1,R,T}}{\|\nabla(u - u_h)\|_{\omega_T} + \varepsilon^{-1/2}\alpha_T\|u - u_h\|_{\omega_T} + \zeta_{H^1,T}} & \text{for residual est.} \\[2ex] \dfrac{\eta_{H^1,D,T}}{\|\nabla(u - u_h)\|_{\omega_T} + \varepsilon^{-1/2}\|u - u_h\|_{\omega_T} + \zeta_{H^1,T}} & \text{for local problem est.} \end{cases}$$

where $\eta_{H^1,*}$ and $\eta_{H^1,*,T}$ stand either for the residual estimator or the local problem error estimator. As before we predict theoretically

$$q_{\mathrm{up}} \lesssim 1 \quad \text{and} \quad q_{\mathrm{low}} \lesssim 1.$$

The next table lists all values, and Figure 3.6.8 displays them graphically. The results coincide with the predictions.

Mesh	Residual error estimator		Local problem error estimator	
	q_{up}	q_{low}	q_{up}	q_{low}
$\mathcal{T}_1$	0.186	0.571	0.235	0.359
$\mathcal{T}_2$	0.155	1.409	0.364	0.368
$\mathcal{T}_3$	0.128	2.792	0.427	0.420
$\mathcal{T}_4$	0.118	3.352	0.514	0.495
$\mathcal{T}_5$	0.112	3.665	0.590	0.532
$\mathcal{T}_6$	0.106	3.813	0.631	0.550

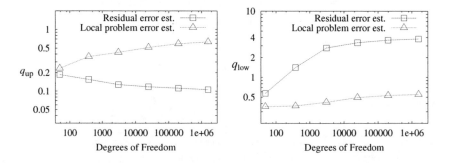

Figure 3.6.8: Left: Upper error bound: $q_{\mathrm{up}} \lesssim 1$
Right: Lower error bound: $q_{\mathrm{low}} \lesssim 1$

Equivalence of the error estimators

Finally we investigate the equivalence of both estimators in the same manner as above. Employing the abbreviations

$$q_{D/R} := \max_{T \in \mathcal{T}_k} \frac{\eta_{H^1,D,T}}{\left(\sum_{T' \subset \omega_T} \eta^2_{H^1,R,T'} \right)^{1/2}} \quad \text{and} \quad q_{R/D} := \max_{T \in \mathcal{T}_k} \frac{\eta_{H^1,R,T}}{\left(\sum_{T' \subset \omega_T} \eta^2_{H^1,D,T'} \right)^{1/2}},$$

Theorem 3.4.3 implies

$$q_{D/R} \lesssim 1 \quad \text{and} \quad q_{R/D} \lesssim 1.$$

This is confirmed by the table below and the related Figure 3.6.9.

Mesh	$q_{D/R}$	$q_{R/D}$
$\mathcal{T}_1$	0.401	1.121
$\mathcal{T}_2$	0.397	4.166
$\mathcal{T}_3$	0.718	4.888
$\mathcal{T}_4$	0.755	4.967
$\mathcal{T}_5$	0.724	5.026
$\mathcal{T}_6$	0.712	5.073

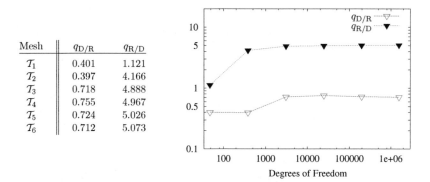

Figure 3.6.9: H^1 seminorm: Equivalence of error estimators: $q_{D/R} \lesssim 1$ and $q_{R/D} \lesssim 1$

Chapter 4

Singularly perturbed convection diffusion problem

The material of the whole chapter originates from [Kun03]. The scientific relation to other chapters and the connection in content is clearly given by the anisotropic behaviour of the solution. In particular we treat a more general problem than in Chapter 3. Therefore also a similar kind of investigation is suitable, with analogous auxiliary lemmas.

To avoid any confusion and misunderstanding we stress that the notation here is similar to that of Chapter 3. Of course some of the terms now get a modified meaning (because of the different PDE). Hence the reader should take care to always take the definitions from this chapter. For example this concerns the energy norm $\|\|v\|\|$, the exact element residual R_T, the residual scaling factor α_T, the error estimators $\eta_{\varepsilon,*,T}$, the local space V_T and the squeezing parameter γ_E.

4.1 Problem description

This chapter deals with the singularly perturbed diffusion–convection–reaction problem

$$\left.\begin{array}{rcll} -\varepsilon\Delta u + \underline{b}\cdot\nabla u + cu & = & f & \text{in } \Omega \\ u & = & 0 & \text{on } \Gamma_D \\ \varepsilon\partial u/\partial n & = & g & \text{on } \Gamma_N. \end{array}\right\} \tag{4.1}$$

Such problems arise e.g. when linearising the Navier–Stokes equations. The convection dominated case is particularly interesting, and thus the diffusion parameter is supposed to be small, $0 < \varepsilon \ll 1$. As a consequence, the solution u of (4.1) frequently has exponential (or regular) boundary layers of width $\mathcal{O}(\varepsilon|\ln\varepsilon|)$, or parabolic (or characteristic) boundary layers of width $\mathcal{O}(\sqrt{\varepsilon}|\ln\sqrt{\varepsilon}|)$, cf. also [RST96, MOS96, Mor96] and the citations therein. Thus the solution u has *anisotropic* features. Later we will discuss the implications for the discretization.

Let us start with the PDE (4.1). Assume that $\Omega \subset \mathbb{R}^d, d = 2, 3$, is a bounded polyhedral domain with Lipschitz boundary $\partial\Omega = \Gamma_D \cup \Gamma_N$, with $\text{meas}_{d-1}(\Gamma_D) > 0$. In our exposition we will only address the more technical 3D case; the 2D analogues can be derived easily.

We are interested in the convection dominated case of (4.1) and demand three standard assumptions.

(A1) $\underline{b} \in W^{1,\infty}(\Omega)^d$, $c \in L_\infty(\Omega)$

(A2) There exists a positive constant c_0: $-\frac{1}{2}\nabla\cdot\underline{b} + c \geq c_0 > 0$

(A3) $\underline{b}\cdot n \geq 0$ on Γ_N (i.e. Neumann boundary conditions only on outflow boundary).

Next, define an *energy norm* which is closely related to the differential equation by

$$\|\!|v|\!\|_\omega^2 := \varepsilon \|\nabla v\|_\omega^2 + c_0 \|v\|_\omega^2 \qquad . \tag{4.2}$$

The appropriate variational formulation corresponding to (4.1) becomes:

$$\text{Find } u \in H_0^1(\Omega): \qquad B(u,v) = \langle F, v \rangle \qquad \forall\, v \in H_0^1(\Omega) \tag{4.3}$$

with
$$\begin{aligned} B(u,v) &:= \varepsilon(\nabla u, \nabla v) + (\underline{b} \cdot \nabla u, v) + (cu, v) \\ \langle F, v \rangle &:= (f, v) + (g, v)_{\Gamma_N} \qquad . \end{aligned}$$

Thanks to assumptions (A1)–(A3) the bilinear form $B(\cdot, \cdot)$ is elliptic and continuous,

$$B(v,v) \geq \|\!|v|\!\|^2 \qquad \forall\, v \in H_0^1(\Omega) \tag{4.4}$$

$$B(v,w)\big|_\omega \leq \|\!|v|\!\|_\omega \left(\max\{1, c_0^{-1} \|c\|_{\infty,\omega}\} \|\!|w|\!\|_\omega + \varepsilon^{-1/2} \|\underline{b}\|_{\infty,\omega} \|w\|_\omega \right) \tag{4.5}$$

$$\forall\, v, w \in H^1(\omega).$$

The Lax Milgram lemma ensures existence and uniqueness of the weak solution u of (4.3).

It is important to note the *gap* between the *ellipticity* of the bilinear form described by (4.4), and the *continuity* given by (4.5). This gap is essentially driven by the factor $\varepsilon^{-1/2} \|\underline{b}\|_{\infty,\omega}$ on the right-hand side of (4.5). Eventually this becomes the major source of difficulties that later prevents uniform efficiency of the estimators. For more details we refer to the sections below.

Let us now turn to the *discretization* of (4.3). Firstly, the aforementioned layers of the exact solution u have strong *anisotropic* behaviour, i.e. they exhibit lower–dimensional features. Problems with anisotropic solutions can be favourably resolved using *anisotropic meshes*. By this we mean meshes with elements whose aspect ratio is not bounded, as in the conventional theory, but can be arbitrarily large. For further reference we refer to Apel [Ape99]; see e.g. also [Kun99, HL98]. As a consequence of using anisotropic elements, the whole theory of *a priori* and *a posteriori* error estimators has to be reinvestigated since the large aspect ratio influences the error bounds adversely. Let us also refer to the extensive literature about convergence on Shishkin meshes (and similar meshes), see [ST97, LS01, ST01, Lin03] for example.

Secondly, standard numerical methods such as the Finite Element Method (FEM) or the Finite Difference Method fail for small $\varepsilon \ll 1$ (unless the discretization is considerably finer than the layer width which is unrealistic for real-world 2D and 3D applications). The resulting discrete 'solution' would feature nonphysical oscillations, and wouldn't have any resemblance with the analytic solution. Figure 4.1.1 impressively illustrates such a situation.

One possible remedy involves additional stabilization. The most successful approaches are the *streamline upwind Petrov Galerkin* method (SUPG), also known as *streamline diffusion finite element method* (SDFEM), the *Galerkin least squares* approximation (GLS), and the *Douglas–Wang* method. Among the extensive literature, we refer to the overview work of [RST96] and [Mor96], and to [HL98, KLR02] for a unified presentation of stabilized Galerkin methods. Furthermore so-called *shock capturing* gained popularity, see e.g. [KLR02, SE00].

Next, introduce a family $\mathcal{F} = \{\mathcal{T}_h\}$ of triangulations $\mathcal{T}_h$ of Ω that consist of tetrahedra. We assume an admissible triangulation in the sense of [Cia78]. Let $V_{0,h} \subset H_0^1(\Omega)$ be the space of continuous, piecewise linear functions over $\mathcal{T}_h$ that vanish on Γ_D.

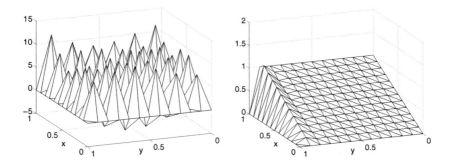

Figure 4.1.1: The necessity of stabilization for the example $-\varepsilon\Delta u + u_y = 0$ with exact solution $u = x(1 - \exp(\frac{y-1}{\varepsilon}))$, $\varepsilon = 10^{-4}$.
Left: Useless discrete solution without stabilization: $\|u_h\|_{\infty,\Omega} \gg \|u\|_{\infty,\Omega}$.
Right: Accurate discrete solution with stabilization.

In this work we are solely concerned with the *streamline upwind Petrov Galerkin* variant of the finite element method. The stabilized discrete variational problem corresponding to (4.3) then reads

$$\text{Find } u_h \in V_{0,h}: \qquad B_\delta(u_h, v_h) = \langle F_\delta, v_h \rangle \qquad \forall v \in V_{0,h} \tag{4.6}$$

with
$$B_\delta(u_h, v_h) := B(u_h, v_h) + \sum_{T \in \mathcal{T}_h} \delta_T \left(-\varepsilon\Delta u_h + \underline{b} \cdot \nabla u_h + c u_h, \underline{b} \cdot \nabla v_h\right)_T$$
$$\langle F_\delta, v_h \rangle := \langle F, v_h \rangle + \sum_{T \in \mathcal{T}_h} \delta_T \left(f, \underline{b} \cdot \nabla v_h\right)_T \qquad .$$

When the real stabilization parameters δ_T vanish for all elements T then (4.6) coincides with the standard Galerkin discretization. In the singularly perturbed case ($\varepsilon \ll 1$) this discretization is unsuitable since it suffers from severe instabilities. Then the stabilized SUPG discretization (corresponding to $\delta_T > 0$) is more appropriate.

The discrete solution u_h exists and is unique if the stabilization parameters δ_T are sufficiently small. Following [RST96, Section III.3.2.1], one can show that

$$0 \le \delta_T \le \frac{1}{2}\min\{c_0\|c\|_{\infty,T}^{-2}, \ h_{min,T}^2 \varepsilon^{-1}\mu^{-2}\}$$

is sufficient (see also [HL98, Section 2.2]). The constant μ is such that the inverse inequality

$$\|\operatorname{div}\nabla v_h\|_T \le \mu h_{min,T}^{-1}\|\nabla v_h\|_T$$

holds for all $v_h \in V_{0,h}$. For the case of piecewise linear functions in $V_{0,h}$ as considered here, this simplifies to $\mu = 0$ and (after some refined calculation) to $0 \le \delta_T \le c_0\|c\|_{\infty,T}^{-2}$. In the sequel we always assume

$$\delta_T \lesssim h_{min,T}\|\underline{b}\|_{\infty,T}^{-1} \qquad \forall T \in \mathcal{T}_h$$

This demand is met for all kinds of stabilization employed in our work, see the numerical experiments of Section 4.4 and [Kun03, Section 6.4]. We remark that the role of stabilization is not completely clear yet, in particular for anisotropic discretizations.

Furthermore it is advantageous to define a so–called *local mesh Peclet number* by

$$\mathrm{Pe}_T := \frac{\|\underline{b}\|_{\infty,T} h_{min,T}}{2\varepsilon} \quad , \tag{4.7}$$

cf. [HL98, Section 2.4] or the isotropic counterpart [RST96, Section III.3.2.1]. This mesh Peclet number relates the ratio of local convection and diffusion to the minimal local mesh size. Small mesh Peclet numbers $\mathrm{Pe}_T \lesssim 1$ will be advantageous while large values $\mathrm{Pe}_T \gg 1$ will cause problems (both theoretically and in practice).

Our main interest is now again in reliable and efficient *a posteriori* error estimators. This topic is by now well understood for symmetric, elliptic PDEs where tight upper and lower error bounds are achieved, cf. the overview work of [Ver96, AO00]. For the *convection dominated* case as considered here, the theory is much less mature. This is mainly caused by the large convection which implies a gap between the ellipticity constant and the boundedness constant of the bilinear form associated with the PDE. Hence the constants in the error bounds depend on the problem parameters. Reliability and/or efficiency of the error estimator may be affected adversely.

The last decade has seen much effort to diminish the influence on the problem data. Angermann [Ang95] was the first to eliminate this dependence completely. Unfortunately the error is measured there in a complicated norm which is defined implicitly via an infinite dimensional variational problem. Hence tight error bounds are achieved on the expense of a norm that is difficult to evaluate.

A more feasible approach is presented by [Ver98a, Ber02] and Kay/Silvester [KS01] which measure the error in the energy norm and the H^1 seminorm, respectively. In all cases the error estimator is *reliable*, i.e. an upper error bound holds. The *efficiency* is associated with the lower error bound and depends on a local mesh Peclet number, cf. below for details.

Formaggia at al. [FPZ01] propose a post–processing based estimator for error functionals. Recently Sangalli [San01] obtained an error estimate in a particular norm for the so–called residual free bubble method. Also worth mentioning are the numerical study of *a posteriori* error estimators by John [Joh00] and the comparisons in [PV00]. They treat several refinement strategies and/or error estimators that are either heuristically derived or mathematically analysed. There valuable conclusions are obtained about reliability of estimators and their suitability for adaptive algorithms. One (slightly surprising) observation is that parabolic layers can be more difficult to treat than exponential layers.

Finally, a maximum norm *a posteriori* estimator for a 1D problem has been proposed recently in [Kop01]. We also mention the vast literature on *a priori* estimates, cf. [Ape99, RST96] and the citations therein.

Here we are chiefly concerned about *a posteriori* estimators for *anisotropic discretizations*. There has been some development in recent years (for several PDEs); exemplarily we mention [Sie96, Kun00, DGP99, Kun01f]. There is a common feature of all those estimators that is different to the isotropic theory. Namely, the reliability and efficiency of the error estimator is not achieved for *arbitrary* anisotropic meshes (i.e. independent of the anisotropic solution). If the anisotropy of the mesh and of the solution are well aligned then tight upper and lower error bounds are obtained. Otherwise there can be an arbitrarily large gap between both error bounds, and the error estimator would be useless. For more details see Section 3.3.

In this chapter we are interested in *a posteriori* error estimators that are suitable for singularly perturbed convection diffusion problems on *anisotropic meshes*. To our knowledge the first such estimator has been proposed by [FPZ01]. They estimate a functional (and not a norm) of the error by means of a related dual solution, *a priori* interpolation estimates, and a postprocessing procedure. This approach suffers from the drawback that only an upper error bound is obtained but not a lower error bound. Hence it is not clear if the error estimator is close to the true error.

In contrast, both of our novel error estimators provide *upper and lower* error bounds. They are tight in certain, well defined circumstances. Our proposals are inspired by (isotropic) estimators for convection diffusion problems [KS01, Ver98a] and by estimators for anisotropic elements [Kun01e, Kun01f]. It turns out that the *upper error bound* depends on the alignment of the anisotropic mesh and of the anisotropic solution. This dependence enters in the same way as for the Poisson equation or reaction diffusion equations, cf. [Kun00, Kun01f]. The *lower error bound* involves a local mesh Peclet number Pe_T. It implies *efficiency* of the error estimator if the mesh Peclet number is small ($\mathrm{Pe}_T \lesssim 1$). This is an result analogous to that of isotropic elements.

The remainder of the chapter is organised as follows. Section 4.2 presents the *residual error estimator*. Afterwards Section 4.3 is devoted to a *local problem error estimator*. Several numerical experiments that illustrate the theoretical predictions are given in Section 4.4. All material stems from [Kun03].

4.2 Residual error estimator

The general methodology of residual error estimation is similar to that of the reaction diffusion problem, cf. the introduction to Section 3.3. Basically the error estimator is defined by measuring and weighting the residuals. The main task is to find appropriate anisotropic weights such that tight upper and lower error bounds can be proven. This requires careful tuning of all ingredients of the proofs of the upper and lower error bounds.

Let us start with the definition of the error estimator. Parts of the theory require residual terms from a finite dimensional space. Therefore we utilize approximations of the exact residuals. In particular replace $g \in L^2(\Gamma_\mathrm{N})$ by g_h which is piecewise constant over the Neumann faces.

Definition 4.2.1 (Element and face residual) *The exact element residual over an element T is given by*

$$R_T := f - (-\varepsilon \Delta u_h + \underline{b} \cdot \nabla u_h + c u_h) \qquad on \ T.$$

The (approximate) element residual r_T is any approximation to R_T that is constant on T, i.e.

$$r_T \in \mathbb{P}^0(T). \tag{4.8}$$

For a face E define the (approximate) face residual $r_E \in \mathbb{P}^0(E)$ by

$$r_E := \begin{cases} \varepsilon \cdot [\![\nabla u_h \, n_E]\!]_E & \text{if } E \subset \Omega \setminus \Gamma \\ g_h - \varepsilon \cdot \partial u_h / \partial n & \text{if } E \subset \Gamma_\mathrm{N} \\ 0 & \text{if } E \subset \Gamma_\mathrm{D} \ . \end{cases}$$

Again $n_E \perp E$ is one of the two unitary normal vectors whereas $n \perp E \subset \Gamma_\mathrm{N}$ denotes the outer unitary normal vector, cf. Chapter 2.

The face residual is also known as gradient jump *or* jump residual*. Note again that the element residual r_T is clearly related to the strong form of the differential equation.*

Similar to Definition 3.3.2 a residual scaling factor will be defined.

Definition 4.2.2 (Residual scaling factor) *For an element T, set*

$$\alpha_T := \min\{c_0^{-1/2}, \varepsilon^{-1/2} h_{min,T}\} \quad . \tag{4.9}$$

Definition 4.2.3 (Residual error estimator) *For a tetrahedron T, the residual error estimator $\eta_{\varepsilon,R,T}$ and the approximation term $\zeta_{\varepsilon,T}$ are defined by*

$$\eta_{\varepsilon,R,T}^2 := \alpha_T^2 \cdot \|r_T\|_T^2 + \varepsilon^{-1/2} \alpha_T \sum_{E \subset \partial T \setminus \Gamma_D} \|r_E\|_E^2$$

$$\zeta_{\varepsilon,T}^2 := \alpha_T^2 \cdot \sum_{T' \subset \omega_T} \|r_{T'} - R_{T'}\|_{T'}^2 + \varepsilon^{-1/2} \alpha_T \sum_{E \subset \partial T \cap \Gamma_N} \|g - g_h\|_E^2 \quad .$$

Furthermore define the corresponding global *expressions by*

$$\eta_{\varepsilon,R}^2 := \sum_{T \in \mathcal{T}_h} \eta_{\varepsilon,R,T}^2 \quad \text{and} \quad \zeta_\varepsilon^2 := \sum_{T \in \mathcal{T}_h} \zeta_{\varepsilon,T}^2 \quad .$$

For convenience, define the mesh Peclet number on the patch ω_T by

$$\mathrm{Pe}_{\omega_T} := \max_{T' \subset \omega_T} \mathrm{Pe}_{T'} \quad . \tag{4.10}$$

Then the main theoretical result can be formulated as follows.

Theorem 4.2.4 (Residual error estimation) *The error is bounded locally from below for all $T \in \mathcal{T}_h$ by*

$$\eta_{\varepsilon,R,T} \lesssim \|\|u - u_h\|\|_{\omega_T} \cdot \left(\max\{1, c_0^{-1}\|c\|_{\infty,\omega_T}\} + \varepsilon^{-1/2} \alpha_T \|\underline{b}\|_{\infty,\omega_T}\right) + \zeta_{\varepsilon,T} \quad . \tag{4.11}$$

Recalling Pe_{ω_T} from (4.10), this lower bound can be rewritten in the slightly weaker form

$$\eta_{\varepsilon,R,T} \lesssim \|\|u - u_h\|\|_{\omega_T} \cdot \left(\max\{1, c_0^{-1}\|c\|_{\infty,\omega_T}\} + \mathrm{Pe}_{\omega_T}\right) + \zeta_{\varepsilon,T} \quad . \tag{4.12}$$

Assume further that the stabilization parameters satisfy $\delta_T \lesssim h_{min,T}/\|\underline{b}\|_{\infty,T}$. Then the error is bounded globally from above by

$$\|\|u - u_h\|\| \lesssim m_1(u - u_h, \mathcal{T}_h) \cdot \left[\eta_{\varepsilon,R}^2 + \zeta_\varepsilon^2\right]^{1/2} \quad . \tag{4.13}$$

Key ideas of the proof: The proof is given in [Kun03].

The derivation of the *lower error bound* (4.11) hinges on bubble functions (both standard and modified ones), and corresponding inverse inequalities. The factor Pe_{ω_T} on the right-hand side of (4.12) is introduced basically by the continuity (4.5) of the bilinear form. Stabilization is no issue here.

The *upper error bound* (4.13) relies on the Galerkin orthogonality and Clément interpolation error estimates. The stabilization requires separate investigation via an inverse inequality for a particular Clément interpolant, and the stability of the Clément interpolation operator. Furthermore the stabilization parameter has to be small enough which is frequently satisfied in practice. ∎

Remark 4.2.5 (Reliability and efficiency) When investigating error estimators, one often encounters the terms *reliable* and *efficient*. Up to approximation terms (or higher order terms), they commonly have the meaning

$$\begin{array}{rcl} \text{(global) Reliability} & \Leftrightarrow & \|\|u - u_h\|\| \lesssim \eta_{\varepsilon,R} \\ \text{(local) Efficiency} & \Leftrightarrow & \eta_{\varepsilon,R,T} \lesssim \|\|u - u_h\|\|_{\omega_T} \end{array} \quad ,$$

i.e. they are closely related to (but not identical with) the upper and lower error bounds. Let us compare these definitions with our main result of Theorem 4.2.4. We conclude that the error is *reliable* whenever the alignment measure is small, $m_1(u - u_h, \mathcal{T}_h) \sim 1$. This will be the case when the anisotropic mesh is well adapted to the anisotropic solution.

The *efficiency* requires a careful distinction for convection–diffusion problems. Recall the lower error bound (4.12),

$$\eta_{\varepsilon,\mathrm{R},T} \lesssim \|\|u - u_h\|\|_{\omega_T} \cdot \left\{ \max\{1, c_0^{-1}\|c\|_{\infty,\omega_T}\} + \mathrm{Pe}_{\omega_T} \right\} + \zeta_{\varepsilon,T} \quad .$$

Assume for a moment $c_0 \sim \|c\|_{\infty,\omega_T}$, i.e. the behaviour of the factor on the right–hand side is essentially determined by the mesh Peclet numbers Pe_{ω_T}. Hence the above error bound implies efficiency only for small mesh Peclet numbers $\mathrm{Pe}_{\omega_T} \lesssim 1$. Such small Peclet numbers arise e.g. for (suitable) anisotropic elements in layer regions of exponential type (i.e. layers of width $\mathcal{O}(\varepsilon)$).

Conversely, for large mesh Peclet numbers the efficiency cannot be guaranteed since the error estimator $\eta_{\varepsilon,\mathrm{R},T}$ may be large even when the error $\|\|u - u_h\|\|_{\omega_T}$ is small. Large Peclet numbers $\mathrm{Pe}_{\omega_T} \gg 1$ will usually arise for elements in coarse mesh regions or parabolic layer regions (i.e. layers of width $\mathcal{O}(\sqrt{\varepsilon})$). Numerical comparisons of [Joh00] indicate that the lacking efficiency is mainly a problem inside parabolic layers since the error is usually much larger there than in coarse mesh regions with a smooth solution.

Finally we remark that the partial loss of efficiency is not due to the anisotropic elements but comes from the dominating convection of the problem, cf. also [KS01, Ver98a]. □

Remark 4.2.6 (Higher order approximate residuals) The approximation of the exact residuals in Definition 4.2.1 is piecewise *constant* over the elements and faces, respectively. The extension to higher order approximate residuals is comparatively simple and described in [Kun03, Section 4.2].

The treatment of higher order ansatz functions to define the approximate solution u_h is given there as well. □

4.3 Local problem error estimator

The key idea is to solve the problem locally with a higher accuracy. The difference to the original (piecewise linear) solution serves as error estimator, cf. the textbooks [AO00, Ver96]. In [Ver98a] a local problem error estimator has been derived for the convection diffusion problem on isotropic elements.

On anisotropic elements, local problem error estimators could be established for the Poisson problem [Kun01a] and a singularly perturbed reaction diffusion problem, cf. Section 3.4 and [Kun01f]. Here we follow the ideas introduced in those works. It is interesting, however, that apparently one has to solve a *reaction diffusion* problem to obtain error bounds for the *convection diffusion* problem (4.1). If a local convection diffusion problem is solved instead then the upper error bound becomes worse. This coincides with the isotropic counterpart.[1]

As before we analyse error bounds when u_h is piecewise linear, and the residuals are approximated by constant values. The general *structure* of the exposition here is quite similar to that of the reaction diffusion problem. Hence the philosophy described in Section 3.4.1 may serve as an adequate motivation again.

---

[1]Note that the first and third error bound of [Ver98a, Proposition 5.1] are not fully correct.

Consider an arbitrary tetrahedron T. We start by defining a local, finite dimensional space V_T that consists of an element bubble function and some squeezed face bubble functions,

$$V_T := \text{span}\{b_T, b_{E, \gamma_E} : E \subset \partial T \setminus \Gamma_D\} \qquad . \tag{4.14}$$

For interior tetrahedra this implies $\dim V_T = 5$. The squeezing parameters γ_E of the squeezed face bubble functions b_{E, γ_E} are now specified to be

$$\gamma_E := \min\left\{1, \frac{h_{min,E}}{h_E}, \frac{\varepsilon^{1/2}}{c_0^{1/2} h_E}\right\}. \tag{4.15}$$

Note that V_T depends implicitly on these parameters γ_E.

The local problem contains a new bilinear form that corresponds to a *reaction diffusion* problem,

$$\tilde{B}(v, w) := \varepsilon(\nabla v, \nabla w) + c_0(v, w) \qquad .$$

This bilinear form is elliptic and continuous, i.e.

$$\tilde{B}(v, v) = |||v|||^2, \qquad \tilde{B}(v, w) \le |||v||| \cdot |||w||| \qquad .$$

The local problem and the error estimator are defined as follows.

Definition 4.3.1 (Local Dirichlet problem error estimator)
Find the unique solution $e_T \in V_T$ of the local variational problem:

$$\tilde{B}(e_T, v_T) = \sum_{T' \subset \omega_T} (r_{T'}, v_T)_{T'} + \sum_{E \subset \partial T \setminus \Gamma_D} (r_E, v_T) \qquad \forall v_T \in V_T \qquad . \tag{4.16}$$

The local and global error estimators then become

$$\eta_{\varepsilon, D, T} := |||e_T|||_{\omega_T} \qquad and \qquad \eta_{\varepsilon, D}^2 := \sum_{T \in \mathcal{T}_h} \eta_{\varepsilon, D, T}^2 \qquad . \tag{4.17}$$

The cornerstone of the analysis is again Lemma 3.4.2. Although this lemma is presented for the *reaction diffusion* problem, it can be evoked here for the *convection diffusion* problem as well since the underlying local space V_T is constructed by the same principles.

Next, the equivalence between the residual error estimator and the local problem error estimator can be established.

Theorem 4.3.2 (Equivalence with residual error estimator) *The local problem error estimator $\eta_{\varepsilon, D, T}$ is equivalent to the residual error estimator $\eta_{\varepsilon, R, T}$ in the following sense:*

$$\eta_{\varepsilon, D, T}^2 \lesssim \sum_{T' \subset \omega_T} \eta_{\varepsilon, R, T'}^2 \tag{4.18}$$

$$\eta_{\varepsilon, R, T}^2 \lesssim \sum_{T' \subset \omega_T} \eta_{\varepsilon, D, T'}^2 \tag{4.19}$$

$$\eta_{\varepsilon, R} \sim \eta_{\varepsilon, D} \qquad . \tag{4.20}$$

All inequalities are uniform in ε.

If T has at least two Neumann boundary faces then the constants in (4.18) and (4.20) can depend on the shape of the Neumann boundary (but do not depend on the triangulation $\mathcal{T}_h$ nor on T).

Key ideas of the proof: The proof which is given in [Kun03, Section 5.1] is very similar to that of Theorem 3.4.3, save for c_0 which need not be 1 here.   ■

Remark 4.3.3 It is important to note that both error estimators depend on the *approximate* residuals r_T and r_E but not on the *exact* residual R_T. In other words, the definition of the error estimator does not depend (explicitly) on the actual PDE.[2]

This explains the close similarity of the proofs of Theorem 3.4.3 and Theorem 4.3.2. In both cases the error estimators ultimately depend on the approximate residuals r_T and r_E, on the geometry data of T and ω_T, and on ε and c_0. Despite the different underlying PDEs, the corresponding treatment differs only by c_0 (which equals 1 in Chapter 3).

Furthermore this observation partly explains why we solve a local *reaction diffusion* problem in (4.16), and omit the convection. This choice already guarantees the equivalence of $\eta_{\varepsilon,\mathrm{D},T}$ and $\eta_{\varepsilon,\mathrm{R},T}$. Therefore a local *convection diffusion* problem cannot improve such an *estimator equivalence* substantially. It is conceivable, however, that a local convection diffusion problem could improve the *error equivalence*. This would require an analysis that bypasses the residual error estimator. Our approach [Kun03] does not provide. It is conceivable that the equilibrated residual method [AO00, Gro02] might be extended to problem (4.1).   □

The main result consists in upper and lower error bounds.

Theorem 4.3.4 (Local Problem error estimation) *The error is bounded locally from below for all $T \in \mathcal{T}_h$ by*

$$\eta_{\varepsilon,\mathrm{D},T} \lesssim \||u - u_h\||_{\omega_T} \cdot \left(\max\{1, c_0^{-1}\|c\|_{\infty,\omega_T}\} + \varepsilon^{-1/2}\alpha_T\|\underline{b}\|_{\infty,\omega_T}\right) + \zeta_{\varepsilon,T} \qquad . \qquad (4.21)$$

This lower bound can be rewritten again in the slightly weaker form

$$\eta_{\varepsilon,\mathrm{D},T} \lesssim \||u - u_h\||_{\omega_T} \cdot \left(\max\{1, c_0^{-1}\|c\|_{\infty,\omega_T}\} + \mathrm{Pe}_{\omega_T}\right) + \zeta_{\varepsilon,T} \qquad .$$

Assume further that the stabilization parameters satisfy $\delta_T \lesssim h_{min,T}/\|\underline{b}\|_{\infty,T}$. Then the error is bounded globally from above by

$$\||u - u_h\|| \lesssim m_1(u - u_h, \mathcal{T}_h) \cdot \left[\eta_{\varepsilon,\mathrm{D}}^2 + \zeta_\varepsilon^2\right]^{1/2} \qquad . \qquad (4.22)$$

If T has at least two Neumann boundary faces then the constant in (4.21) can depend on the shape of the Neumann boundary (but not on T or $\mathcal{T}_h$).

Key ideas of the proof: The proof can be found in [Kun03].

The upper error bound (4.22) follows from the equivalence (4.20) of $\eta_{\varepsilon,\mathrm{D}}$ with the residual error estimator $\eta_{\varepsilon,\mathrm{R}}$, and the error bound (4.13) for the latter estimator.

The derivation of the lower error bound (4.21) employs a reformulation of the local problem (4.16) via partial integration, combined with Lemma 3.4.2. The additional factor on the right-hand side of (4.21) stems from the continuity (4.5) of the bilinear form.

Note again the similarity to the error bounds of the residual estimator, as given by Theorem 4.2.4.   ■

---

[2]In order to avoid any misunderstanding: The *proof* of the error bounds relies heavily on the governing PDE. Hence each error estimator has to be tailored carefully for each PDE, even if this is not immediately evident from the definition.

Remark 4.3.5 (Higher order approximate residuals) One may try to extend our local problem error estimation procedure to cover higher order residuals or higher order ansatz functions for u_h as well. Unfortunately this poses considerable technicalities for the analysis and requires increased numerical effort for the implementation, as described in [Kun03, Section 5.2]. This observation is in contrast to the residual error estimator whose extension to higher order residuals is (more or less) straight forward, cf. the previous section. $\square$

4.4   Numerical experiments

Among the many numerical experiments we present three examples; they underline and confirm the theoretical predictions. The whole material is taken from [Kun03].

- The first and easiest example features a convection diffusion problem with *vanishing convection* (i.e. the simpler reaction diffusion problem). Although this example does not feature 'dominating convection', we have included it since it allows us to distinguish between effects that are due to the anisotropic discretization, and effects that are caused by a large convection. The comparison with the next two examples is particularly interesting and informative.

- The second example describes a convection diffusion problem with *exponential boundary layers*.

- The third example is the most difficult one; the underlying convection diffusion equation gives rise to a *parabolic boundary layer*.

We present the main theoretical results for the *residual error estimator* $\eta_{\varepsilon,\mathrm{R}}$ alone, cf. Theorem 4.2.4 of Section 4.2. In all examples we consider four values of the perturbation parameter, namely $\varepsilon = 10^{-1}, 10^{-2}, 10^{-3}, 10^{-6}$. The methodology to investigate the quality of the solution process and of the error estimation is described in detail in Section 4.4.1 and repeated subsequently. Furthermore we recommend the comparison with Section 3.6 where the results for the reaction diffusion problem have been presented. Additional remarks (e.g. about the choice of the stabilization parameters δ_T) can be found in [Kun03, Section 6].

4.4.1   Example 1

The reaction diffusion problem described here is the special case of the convection diffusion problem (4.1) with vanishing convection. In conjunction with the subsequent examples this allows to separate the effects caused by a dominating convection.

In this example the local mesh Peclet number vanishes; $\mathrm{Pe}_T = 0$ for all elements $T \in \mathcal{T}_h$. This favourite property implies that the corresponding variational problem and the discrete problem are symmetric, respectively, and can be solved without stabilization ($\delta_T = 0$ for all $T \in \mathcal{T}_h$). Note that similar investigations in 3D can be found in [Kun01e, Kun01f] and Section 3.6; for 2D results see [HL98, Ex. 4.2]. Here we have included this example to have a comparison with problems with non–vanishing convection.

With $\underline{b} = (0,0)^\top$ and $c = c_0 = 1$ the PDE becomes

$$-\varepsilon \Delta u + u \;=\; 0 \qquad \text{in } \Omega = (0,1)^2 \quad .$$

The exact solution is prescribed to be

$$u := \mathrm{e}^{-x/\sqrt{\varepsilon}} + \mathrm{e}^{-y/\sqrt{\varepsilon}}$$

and exhibits exponential boundary layers of width $\mathcal{O}(\sqrt{\varepsilon}\,|\ln\varepsilon|)$ along the lines $x = 0$ and $y = 0$. The Dirichlet boundary data on $\Gamma_{\mathrm{D}} := \partial\Omega$ are set accordingly.

We utilize a sequence of three–directional triangular Shishkin type meshes. More precisely, the 2D mesh is the tensor product of two 1D Shishkin type meshes with transition point $\tau := \min\{1/2, \sqrt{\varepsilon}\,|\ln\varepsilon|\}$.

Figure 4.4.2 presents the decrease of the error in the energy norm. The optimal rate of convergence of approximately $\|\!|u - u_h|\!\| = \mathcal{O}(\mathrm{DoF}^{-0.5})$ confirms that the chosen meshes are appropriate to resolve the boundary layers. Judging from our experience in [Kun99, Kun01e] and Section 3.6 we expect the alignment measure $m_1(u - u_h, \mathcal{T}_h)$ to be of moderate size, i.e. in the range of $2 \ldots 4$. Additionally we display the global residual error estimator $\eta_{\varepsilon,\mathrm{R}}$ which always overestimates the error by a factor of about 5.

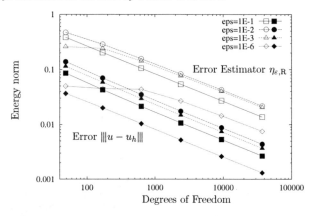

Figure 4.4.2: Error $\|\!|u - u_h|\!\|$ (filled symbols)
Residual error estimator $\eta_{\varepsilon,\mathrm{R}}$ (empty symbols)

Next we investigate the main theoretical results which are the upper and lower error bounds of Theorem 4.2.4. In order to present the underlying inequalities (4.13) and (4.11) appropriately, we reformulate them by defining the ratios of left–hand side and right–hand side, respectively:

$$q_{\mathrm{up}} := \frac{\|\!|u - u_h|\!\|}{\left[\eta_{\varepsilon,\mathrm{R}}^2 + \zeta_\varepsilon^2\right]^{1/2}}$$

$$q_{\mathrm{low}} := \max_{T \in \mathcal{T}_h} \frac{\eta_{\varepsilon,\mathrm{R},T}}{\|\!|u - u_h|\!\|_{\omega_T} \cdot \left(\max\{1, c_0^{-1}\|c\|_{\infty,\omega_T}\} + \varepsilon^{-1/2}\alpha_T\|\underline{b}\|_{\infty,\omega_T}\right) + \zeta_{\varepsilon,T}}.$$

The first ratio q_{up} (or its inverse) is frequently referred to as *effectivity index* and measures the *reliability* of the estimator. The second ratio is related to the *efficiency* of the estimator.

Note further that the factor in the denominator of q_{low} simplifies to

$$\max\{1, c_0^{-1}\|c\|_{\infty,\omega_T}\} + \varepsilon^{-1/2}\alpha_T\|\underline{b}\|_{\infty,\omega_T} \equiv 1$$

for this first example since $\underline{b} = (0,0)^{\mathsf{T}}$. Additionally the approximation terms vanish, $\zeta_{\varepsilon,T} = \zeta_\varepsilon = 0$.

The lower and upper error bound (4.11), (4.13) now correspond to

$$q_{\text{low}} \lesssim 1 \qquad \text{and} \qquad q_{\text{up}} \lesssim m_1(u - u_h, T_h) \qquad .$$

In the right part of Figure 4.4.3 we observe indeed that q_{low} is bounded from above by 2.0. Hence the estimator is *efficient*.

In order to investigate the upper error bound, recall first that the alignment measure $m_1(u - u_h, T_h)$ is expected to be of moderate size $(2 \dots 4)$ since we employ well adapted meshes. Hence the corresponding ratio q_{up} should be bounded from above which is confirmed by the experiment (left part of Figure 4.4.3). As soon as a reasonable resolution of the layer is achieved, the quality of the upper error bound is independent of ε. Thus the estimator is also *reliable*.

Finally we note that the qualitative and the quantitative behaviour of the error estimator is very similar to the 3D counterpart, as described in Section 3.6 and [Kun01e].

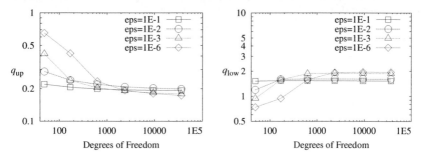

Figure 4.4.3: Left: Upper error bound: $q_{\text{up}} \lesssim 1$
Right: Lower error bound: $q_{\text{low}} \lesssim 1$

4.4.2 Example 2

The second example is taken from [Lin00] and constitutes a typical convection–diffusion problem. It involves convection along $\underline{b} = (b_1, b_2)^\top = (2, 3)^\top$, and we set $c = c_0 = 1$. For the resulting PDE

$$-\varepsilon \Delta u + 2u_x + 3u_y + u = f \qquad \text{in } \Omega = (0,1)^2$$

we prescribe the analytical solution

$$u = \sin(x)(1 - e^{-2(1-x)/\varepsilon}) \cdot y^2 (1 - e^{-3(1-y)/\varepsilon}) \qquad .$$

The right–hand side f and the Dirichlet boundary data on $\partial\Omega$ are chosen accordingly.

The solution shows exponential boundary layers along the outflow boundary at $x = 1$ and $y = 1$. This choice of u serves as a typical example for boundary layers that are caused by incompatibilities of f and the boundary data.

Similar to the previous example we employ a three–directional 2D Shishkin mesh. With N denoting the number of nodal points in x and y direction, the mesh transition points are placed at

$$\tau_x := \min\{1/2, 2\varepsilon \ln N/b_1\} \qquad , \qquad \tau_y := \min\{1/2, 2\varepsilon \ln N/b_2\} \qquad ,$$

respectively, cf. [LS01, Section 2.2]. Hence inside the layer region the local mesh Peclet number is small, $Pe_T \sim 1$.

The unsymmetric variational problem requires stabilization to yield accurate results. For the stabilization parameter δ_T we follow the isotropic proposal in [Ver98a] and define the anisotropic counterpart by

$$\delta_T := \frac{h_{min,T}}{2\|\underline{b}\|_{\infty,T}} \cdot (\coth(Pe_T) - Pe_T^{-1}) \quad .$$

This implies little stabilization in the layer region and comparatively large stabilization in the coarse mesh region.

The results are presented in a similar fashion as for example 1. Start with the decrease of the error $\|u - u_h\|$ and the error estimator, respectively, which is depicted in Figure 4.4.4. The convergence rate is approximately $\|u - u_h\| = \mathcal{O}(DoF^{-0.41})$ and thus (most likely) sub–optimal. The error estimator $\eta_{\varepsilon,R}$ overestimates the error by a factor of approximately 5 but behaves similarly otherwise.

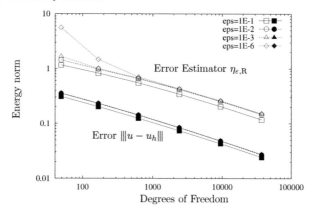

Figure 4.4.4: Error $\|u - u_h\|$ (filled symbols)
Residual error estimator $\eta_{\varepsilon,R}$ (empty symbols)

In order to assess the error bounds, we compute again the ratios q_{low} and q_{up} and present them in Figure 4.4.5, cf. the previous example. Starting with the lower error bound (4.11), we observe that q_{low} is very moderately growing, and appears to be bounded by 5. This coincides with the theoretical predictions (and other experiments). Note, however, that the right–hand side of (4.11) now contains the factor $\max\{1, c_0^{-1}\|c\|_{\infty,\omega_T}\} + \varepsilon^{-1/2}\alpha_T\|\underline{b}\|_{\infty,\omega_T}$. This additional factor is of order 1 for small mesh Peclet numbers $Pe_T \lesssim 1$, i.e. for elements inside the layer region. Then the error estimator is *efficient*.

Conversely, the aforementioned factor becomes large for elements with a large mesh Peclet number $Pe_T \gg 1$, i.e. in the coarse mesh region. Numerical investigation strongly suggests that this factor cannot be omitted. As a consequence the efficiency of the error estimator deteriorates as Pe_T becomes large. On the other hand this may not be too much of a disadvantage since the mesh Peclet number should be large only in regions where the solution u is smooth and the error is already small.

When investigating the upper error bound (4.13), the corresponding ratio q_{up} is bounded by the alignment measure $m_1(u - u_h, \mathcal{T}_h)$. Again we expect m_1 to be of moderate size (say

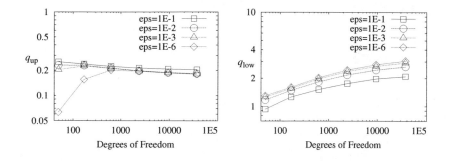

Figure 4.4.5: Left: Upper error bound: $q_{up} \lesssim 1$
Right: Lower error bound: $q_{low} \lesssim 1$

$2\ldots 4$). Hence q_{up} has to bounded from above which is confirmed by the left part of Figure 4.4.5. Consequently the error estimator is *reliable*.

4.4.3   Example 3

This example features a so–called parabolic layer. The numerical comparison of [Joh00] reveals that such layers are much more difficult to treat than exponential layers, in particular when designing adaptive algorithms. The theoretical knowledge is also less developed although parabolic layers may be equally important in practical applications. The difficulties become apparent in the experiment below.

Our test is largely inspired by [HL98, Example 4.3]. With $\underline{b} = (1,0)^\top$ and $c = c_0 = 1$ the PDE becomes

$$-\varepsilon\Delta u + u_x + u = f \qquad \text{in } \Omega = (0,1)^2 \qquad .$$

The exact solution is prescribed to be

$$u = \frac{1}{\sqrt{1+x}} \cdot \exp\left(-\frac{y^2}{4\varepsilon(1+x)}\right)$$

with an appropriate right–hand side f and the Dirichlet boundary data on $\partial\Omega$. This solution u displays a typical parabolic layer along the line $y = 0$. Note that the layer width of $\mathcal{O}(\sqrt{\varepsilon})$ is much larger than for an exponential layer.

Again a three–directional 2D Shishkin mesh is employed. With N denoting the number of nodal points in y direction, the mesh transition point is set to

$$\tau_y := \min\{1/2, 2\sqrt{\varepsilon}\ln N\}$$

Hence the elements T in the layer region have a minimal dimension $h_{min,T} \sim \sqrt{\varepsilon}$ which is (much) larger than ε (unless the mesh is very fine which is unrealistic for small ε). Consequently the local mesh Peclet number is (much) larger than 1 *even in the critical layer region*. This observation is a key difference to the exponential layers (cf. example 2).

When solving the variational problem, we apply the stabilization proposed in [HL98, Section 3.3] (up to the factor $1/3$), namely

$$\delta_T = \frac{h_{min,T}}{3\|\underline{b}\|_{\infty,T}} \cdot \min\{1, \mathrm{Pe}_T\} \qquad .$$

With this setting the convergence rate of the error $\||u - u_h\||$ in the energy norm is $\mathcal{O}(\mathrm{DoF}^{-0.50})$ for $\varepsilon = 10^{-1}$ and drops to about $\mathcal{O}(\mathrm{DoF}^{-0.40})$ for $\varepsilon = 10^{-6}$. The error estimator $\eta_{\varepsilon,\mathrm{R}}$ behaves qualitatively similarly and overestimates the error as in the previous example.

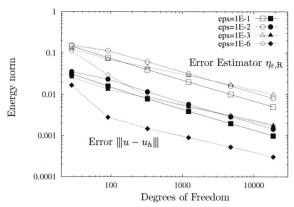

Figure 4.4.6: Error $\||u - u_h\||$ (filled symbols)
Error estimator $\eta_{\varepsilon,\mathrm{R}}$ (empty symbols)

Next we investigate the error bounds in Figure 4.4.7. For the lower error bound (4.11) we compute again q_{low}. In accordance with the theory, q_{low} is bounded from above. Since most values of q_{low} are much smaller than in previous examples, the error bound (4.11) is not sharp here. This may be caused by the parabolic structure of the boundary layer.

A more detailed inspection reveals that the *efficiency* is lost only for coarse mesh elements, i.e. where the mesh Peclet number is large. Inside the parabolic layer the mesh Peclet number is still large (unless the mesh is unrealistically fine for small ε). Nevertheless the numerical results for this example yield that the error estimator is still quite efficient, with only a mild decrease of efficiency for small ε and fine meshes. This behaviour is somewhat better than we can expect from the theory.

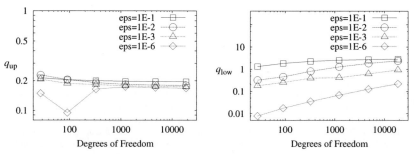

Figure 4.4.7: Left: Upper error bound: $q_{\mathrm{up}} \lesssim 1$
Right: Lower error bound: $q_{\mathrm{low}} \lesssim 1$

The upper error bound is presented in the left part of Figure 4.4.7. This upper bound (4.13) is not influenced by a (large) mesh Peclet number Pe_T, which should be reflected in the numerical behaviour. Indeed we notice the same performance as in the previous example, i.e. q_{up} is bounded from above, and hence the error estimator is *reliable*. No adverse influence on the reliability is seen that could stem from large values of Pe_T.

Chapter 5

The Stokes problem

5.1 Introduction and general remarks

The whole material is taken from [CKN03]. It is a joint work with Prof. S. Nicaise and Dr. E. Creusé, both from Univ. de Valenciennes, France. The material has been developed in close collaboration. Sections 4.1 and 4.2 of [CKN03] have been derived mainly by G. Kunert. Section 5 of [CKN03] has been initiated by S. Nicaise. The numerical experiments of [CKN03, Section 6] have been carried out by E. Creusé. The remaining material has been proposed, investigated and analysed together.

The relationship to the material of the other chapters is given again by the same *anisotropy* of the underlying solution. This leads to a similar style of analysis with similar tools, cf. Chapter 2.

Motivation and state of the art

In this chapter we investigate the Stokes problem. In certain situations the solution has strong directional features, for example edge singularities. Such a so-called *anisotropic solution* occurs e.g. for concave edges in three dimensional domains, cf. [ANS01b]. As in the previous chapters, it is natural to reflect this anisotropy of the solution also by a corresponding anisotropic discretization.

Furthermore we are concerned with *a posteriori* error estimators which are vitally important for adaptive algorithms and quality control. Particular emphasis is given to the Stokes problem in 3D domains since anisotropic solutions arise there generically. In addition we also treat nonconforming discretizations because they are frequently applied and (comparatively) simple to implement. Altogether we strive to derive and analyse *a posteriori* error estimators that are suitable for *anisotropic discretizations*.

There has been much research for *a posteriori* error estimators for isotropic discretizations of the Stokes problem (mainly for 2D domains), cf. [Ver89, BW90, DDP95, AO97, KS00, CF01] to name but a few. On anisotropic meshes, however, these isotropic estimators usually fail since the lower and upper error bound differ by a factor which is (at most) proportional to the aspect ratio of the anisotropic elements. This potentially unbounded factor renders the isotropic, conventional estimators useless. Hence in the last decade there has been increasing research to find adapted estimators for anisotropic meshes, cf. [Sie96, Kun99, Kun00, Kun01e, FPZ01, DGP99, Ran01]. It turns out that the *upper error bound* is the crucial issue which involves a proper alignment of the anisotropic mesh, see e.g. [Kun00]. Thus we may examine the existing approaches to derive upper error bounds for the Stokes problem on *isotropic* meshes. One encounters just a few techniques (which can partially overlap):

- the residual error estimator method for conforming approximations based on the continuous inf-sup condition [Ver89, KS00],
- the residual error estimator method for nonconforming approximation based on the continuous inf-sup condition (applied to the pressure error alone) and on a Helmholtz like decomposition of the error [DDP95, CF01][1],
- the hierarchical basis method based on a saturation assumption [Ban98, Ran01],
- the local residual method which requires the solution of local Stokes problems [Ver89, BW90, KS00], or of local Poisson problem [AO97, JL00],
- error estimation by duality arguments [Joh98, Bec00, Bur01].

Our goal here is to extend the residual error estimator methods to anisotropic meshes in 2D and 3D domains and to both conforming and nonconforming discretizations. We endeavour to identify minimal assumptions on the elements in order to obtain an equivalence between the error norm and the residual error estimator. This approach allows to make a unified analysis and to extend former results obtained for particular elements on isotropic meshes to a large class of elements on isotropic and anisotropic meshes.

All our error estimators that we present below are novel. To our knowledge the exists only one further estimator that is suitable for anisotropic meshes, namely [Ran01]. There 2D anisotropic triangular meshes are treated in conjunction with the $\mathbb{P}^1 - \mathbb{P}^0$ nonconforming Crouzeix-Raviart element. A hierarchical error estimator is proposed. As it is common for these kinds of estimators, the reliability hinges on a saturation assumption. Unfortunately it is not clear how to ensure this assumption on anisotropic meshes. Furthermore the analysis is based on [AO97, Theorem 1.1]. That theorem, however, is given for a conforming method alone; it is not clear if an extension to nonconforming methods is as simple as claimed in [Ran01].

As in the previous chapters we are now mainly interested in reliable and efficient error estimation for anisotropic discretizations. Thus in Section 5.2 we present the problem, the discretization and the assumptions on it, and some notation. Section 5.3 is devoted to several finite element pairs that are suitable for anisotropic discretizations. The error estimators and the main error bounds are given in Section 5.4. The numerical experiments of Section 5.5 complement our exposition.

Notational issues

In the course of our investigation it turned out that an adapted and modified notation is more appropriate for the treatment of the Stokes problem. This is mainly due to the following reasons.

- The solution consists of a vector valued velocity and a scalar pressure function. The analysis contains many scalar valued, vector valued, or matrix valued terms, respectively. These differences will be reflected by the notation which improves readability considerably.
- Chapters 3 and 4 treat tetrahedral or triangular elements. In contrast we now consider more elements and a large variety of settings.

---

[1]The paper [Ver91] has a similar aim but the consistency error (which is related to tangential gradient jumps) is omitted although it is not of higher order in general.

In this chapter, vector valued terms are denoted by a single underline, for example $\underline{u}$, $\underline{\text{curl}}\ w$. Matrix valued items are denoted by a double underline, e.g. $\underline{\underline{s}}$, $\underline{\underline{J}}_{E,t}$.

We emphasize that some notation has to be redefined. To avoid any confusion, we mention this changed notation here and give the precise definition later.

- Since we treat more elements than just triangles and tetrahedra, the description for the elements has to be adapted. This concerns the main anisotropy vectors $\underline{p}_{i,T}$, the corresponding anisotropic lengths $h_{i,T}$ and the height $h_{E,T}$.
- A matrix is defined similar to C_T from (2.1) but now denoted by $\underline{\underline{C}}_T$.
- The definition of the alignment measure $m_1(\cdot, \cdot)$ has changed in notation but keeps its meaning otherwise.

Further, the gradient of a *scalar* function is the usual (column) vector. In contrast, the gradient of a *vector* function $\underline{v}$ contains the gradient of its components in a rowwise fashion, i.e. $\nabla \underline{v} := (\partial_{x_j} v_i)_{1 \le i,j \le d}$ (with i being the row index and j being the column index).[2]

5.2 Problem description, discretization, notation

In this chapter we consider the stationary Stokes problem with Dirichlet boundary conditions. Given a vector valued function $\underline{f} \in [L^2(\Omega)]^d$ find a vector function $\underline{u}$ representing the velocity of the fluid, and a scalar function p that describes the pressure, satisfying

$$\left.\begin{array}{rcll} -\Delta \underline{u} + \nabla p & = & \underline{f} & \text{in } \Omega \\ \text{div}\, \underline{u} & = & 0 & \text{in } \Omega \\ \underline{u} & = & 0 & \text{on } \Gamma_{\mathrm{D}} \equiv \partial\Omega. \end{array}\right\} \tag{5.1}$$

5.2.1 Mixed variational formulation and discretization

Let us fix a bounded domain Ω of $\mathbb{R}^d$, $d = 2$ or 3, with a Lipschitz boundary, and consider the Stokes problem (5.1). To obtain the corresponding mixed variational formulation, introduce the velocity space V and the pressure space Q by

$$V = [H_0^1(\Omega)]^d := \{\underline{v} \in [H^1(\Omega)]^d : \underline{v} = \underline{0} \text{ on } \partial\Omega\},$$
$$Q = L_0^2(\Omega) := \{q \in L^2(\Omega) : \int_\Omega q = 0\},$$

and the bilinear forms

$$a(\underline{v}, \underline{w}) := \int_\Omega \nabla \underline{v} : \nabla \underline{w}, \qquad b(\underline{v}, q) := -\int_\Omega q\, \text{div}\, \underline{v} \qquad \forall \underline{v}, \underline{w} \in V, \forall q \in Q.$$

Here $\underline{\underline{A}} : \underline{\underline{B}} := \sum_{i,j=1}^{d} A_{ij} B_{ij}$ stands for the standard contraction of two matrices $\underline{\underline{A}}$ and $\underline{\underline{B}}$.

The mixed formulation then reads: Find $(\underline{u}, p) \in V \times Q$ such that

$$\left.\begin{array}{rcll} a(\underline{u}, \underline{v}) + b(\underline{v}, p) & = & \int_\Omega \underline{f}\, \underline{v} & \forall v \in V, \\ b(\underline{u}, q) & = & 0 & \forall q \in Q. \end{array}\right\} \tag{5.2}$$

Following [GR86, Theorem I.5.1], the weak solution $(\underline{u}, p) \in V \times Q$ exists and is unique.

---

[2]Note that the gradient operator does not carry the vector/matrix notation.

Let us now turn to the discretization of (5.2). Start with the domain Ω which is discretized by a conforming mesh $\mathcal{T}_h$ in the sense of [Cia78]. Consider an approximate velocity space V_{veloc} and an approximate pressure space $Q_{\text{pre}} \subset Q$ both made of certain polynomials on each element T of the triangulation $\mathcal{T}_h$. For $V_{\text{veloc}} \subset V$ the discretization is called *conforming*. Otherwise, for $V_{\text{veloc}} \not\subset V$ one has a *nonconforming* discretization. A precise description of the properties that these approximation spaces V_{veloc} and Q_{pre} have to satisfy is given below. Moreover many examples of suitable spaces are presented in Section 5.3.

Since the velocity approximation space V_{veloc} may not be included in the velocity space V, we define the approximate solution by using the weaker bilinear forms $a_h(.,.)$ and $b_h(.,.)$:

$$a_h(\underline{v},\underline{w}) \quad := \quad \sum_{T \in \mathcal{T}_h} \int_T \nabla \underline{v} : \nabla \underline{w}, \qquad \forall \underline{v}, \underline{w} \in V_{\text{veloc}}, \tag{5.3}$$

$$b_h(\underline{v}, q) \quad := \quad -\sum_{T \in \mathcal{T}_h} \int_T q \operatorname{div} \underline{v}, \qquad \forall \underline{v} \in V_{\text{veloc}}, q \in Q_{\text{pre}}. \tag{5.4}$$

The mixed finite element formulation reads now: Find $(\underline{u}_h, p_h) \in V_{\text{veloc}} \times Q_{\text{pre}}$ such that

$$\left.\begin{array}{rcll} a_h(\underline{u}_h, \underline{v}_h) + b_h(\underline{v}_h, p_h) & = & \int_\Omega \underline{f}\,\underline{v}_h & \forall \underline{v}_h \in V_{\text{veloc}}, \\ b_h(\underline{u}_h, q_h) & = & 0 & \forall q_h \in Q_{\text{pre}}. \end{array}\right\} \tag{5.5}$$

Existence and uniqueness of a discrete solution will be answered below.

If one has $\underline{v}|_T \in [H^1(T)]^d$ for all T in $\mathcal{T}_h$, then we can define a *broken gradient norm* on a subset ω of Ω by :

$$\|\nabla_\mathcal{T} \underline{v}\|_\omega^2 := \sum_{T \subset \omega} \|\nabla \underline{v}\|_T^2.$$

Thus the space V_{veloc} can be equipped with the seminorm $\|\nabla_\mathcal{T} \underline{v}\|_\Omega = a_h^{1/2}(\underline{v}, \underline{v})$.

5.2.2   Elements and redefined notation

Start with the conforming mesh $\mathcal{T}_h$. In 2D, all elements are either triangles or rectangles. In 3D the mesh consists either of tetrahedra, of rectangular hexahedra, or of rectangular pentahedra (i.e. prisms where the triangular faces are orthogonal to the rectangular faces), cf. also the figures below. Because of this large variety of elements we have to redefine some of the previous notation.

Let us start with a description of the elements T and the corresponding reference elements $\bar{T}$. For an element T we redefine 2 or 3 *anisotropy vectors* (or anisotropic directions) $\underline{p}_{i,T}, i = 1 \ldots d$, that reflect the main anisotropy directions of that element. The elements, the reference elements, and the anisotropy vectors are defined and visualised in Table 5.2.1.

The anisotropy vectors $\underline{p}_{i,T}$ are enumerated such that their lengths are decreasing, i.e. $|\underline{p}_{1,T}| \geq |\underline{p}_{2,T}| \geq |\underline{p}_{3,T}|$ in the 3D case, and analogously in 2D. The *anisotropic lengths* of an element T are now redefined by

$$h_{i,T} := |\underline{p}_{i,T}|$$

which implies $h_{1,T} \geq h_{2,T} \geq h_{3,T}$ in 3D. The smallest of these lengths is particularly important; thus recall

$$h_{min,T} := h_{d,T} \equiv \min_{i=1\ldots d} h_{i,T}.$$

Element T	Reference element $\bar{T}$	Anisotropy vectors $\underline{p}_{i,T}$
Triangle	$0 \leq \bar{x}, \bar{y}$ $\bar{x} + \bar{y} \leq 1$	$\underline{p}_{1,T}$ longest edge $\underline{p}_{2,T}$ height vector
Rectangle	$0 \leq \bar{x}, \bar{y} \leq 1$	$\underline{p}_{1,T}$ longest edge $\underline{p}_{2,T}$ height vector
Tetrahedron	$0 \leq \bar{x}, \bar{y}, \bar{z}$ $\bar{x} + \bar{y} + \bar{z} \leq 1$	$\underline{p}_{1,T}$ longest edge $\underline{p}_{2,T}$ height in largest face that contains $\underline{p}_{1,T}$ $\underline{p}_{3,T}$ remaining height
Hexahedron	$0 \leq \bar{x}, \bar{y}, \bar{z} \leq 1$	$\underline{p}_{1,T}$ longest edge $\underline{p}_{2,T}$ height in largest face that contains $\underline{p}_{1,T}$ $\underline{p}_{3,T}$ remaining height
Pentahedron (Prism)	$0 \leq \bar{x}, \bar{y}, \bar{z} \leq 1$ $\bar{x} + \bar{y} \leq 1$	longest edge in triangle; height in triangle; height over triangle (see figure)

Table 5.2.1: Elements, reference elements, and main anisotropy vectors

Finally the anisotropy vectors $\underline{p}_{i,T}$ are arranged columnwise to redefine a matrix

$$
\left.
\begin{aligned}
\underline{\underline{C}}_T &:= [\underline{p}_{1,T}, \underline{p}_{2,T}] \in \mathbb{R}^{2\times 2} && \text{in 2D} \\
\underline{\underline{C}}_T &:= [\underline{p}_{1,T}, \underline{p}_{2,T}, \underline{p}_{3,T}] \in \mathbb{R}^{3\times 3} && \text{in 3D.}
\end{aligned}
\right\}
\tag{5.6}
$$

Note that $\underline{\underline{C}}_T$ is orthogonal since the anisotropy vectors $\underline{p}_{i,T}$ are orthogonal too, and

$$
\underline{\underline{C}}_T^\top \underline{\underline{C}}_T = \operatorname{diag}\{h_{1,T}^2, \ldots, h_{d,T}^2\}.
$$

Furthermore introduce the *height* $h_{E,T}$ over an edge/face E of an element T by

$$
h_{E,T} := \frac{|T|}{|E|} \cdot
\begin{cases}
1 & T \text{ is rectangle or hexahedron} \\
d & T \text{ is triangle or tetrahedron} \\
1 & \text{for triangular face } E \text{ of pentahedron } T \\
2 & \text{for rectangular face } E \text{ of pentahedron } T.
\end{cases}
$$

In 3D, we further need the minimal size $\varrho(E)$ of a rectangular face E, i.e., $\varrho(E)$ is the smallest of the lengths of the edges of E.

Next we have to redefine the *alignment measure* $m_1(\cdot, \cdot)$ since we are dealing with vector (and matrix) valued functions.

Definition 5.2.1 (Alignment measure) *Let $\underline{v} \in [H^1(\Omega)]^d$ be a vector valued function, and $\mathcal{T}_h$ be a triangulation. The* alignment measure $m_1(\,\cdot\,,\,\cdot\,)$ *is then defined by*

$$m_1(\underline{v}, \mathcal{T}_h) := \frac{\left(\sum_{T \in \mathcal{T}_h} h_{min,T}^{-2} \| \nabla \underline{v} \cdot \underline{C}_T \|_T^2 \right)^{1/2}}{\| \nabla \underline{v} \|}. \tag{5.7}$$

For a matrix valued function $\underline{s} = (s_{i,j})_{i,j=1}^3 \in [H^1(\Omega)]^{3 \times 3}$ the definition is componentwise,

$$m_1(\underline{s}, \mathcal{T}_h) := \frac{\left(\sum_{T \in \mathcal{T}_h} h_{min,T}^{-2} \sum_{i,j=1}^3 \|(\nabla s_{i,j})^\top \cdot \underline{C}_T\|_T^2 \right)^{1/2}}{\left(\sum_{T \in \mathcal{T}_h} \sum_{i,j=1}^3 \|\nabla s_{i,j}\|_T^2 \right)^{1/2}}.$$

The main difference to Definition 2.3.1 for a *scalar* function v is that now $\underline{v}$ is a *vector* function. In particular the gradient ∇v of a scalar function is defined as a *column* vector. In contrast, the gradient $\nabla \underline{v}$ of a vector function $\underline{v}$ is a matrix whose *rows* are formed by the gradient of each (scalar) component of $\underline{v}$. Apart from this notational difference, both definitions are the same.

We conclude with some additional notation. The set of all (interior and boundary) edges (2D) or faces (3D) of the triangulation will be denoted by $\mathcal{E}$. In 3D we further use the set of all *rectangular* faces of the triangulation that we shall denote by $\mathcal{E}_\square$.

5.2.3 Requirements on the finite element spaces

In our analysis, a Clément type interpolation operator $I_{Cl}^0 : H^1(\Omega) \to H_0^1(\Omega)$ plays a vital role. In particular we need to specify the image space $\text{Im}(I_{Cl}^0)$. It consists of (globally) continuous functions that vanish on the boundary $\partial\Omega$, and whose representation on each element is as follows:

- linear for triangles or tetrahedra,
- bilinear for rectangles and trilinear for hexahedra,
- the usual restricted bilinear space for pentahedra (i.e. linear on both triangular faces and bilinear on each rectangular face).

General conditions: Two assumptions (G1) and (G2) are necessary.

(G1) The velocity space V_{veloc} is large enough such that it contains the Clément interpolation space, i.e. $[\text{Im}(I_{Cl}^0)]^d \subset [H_0^1(\Omega)]^d \cap V_{\text{veloc}}$.

(G2) In order to obtain robust discrete solutions, the element pairs have to be stable (i.e. the discrete inf-sup condition is satisfied, cf. [AR03]). Note that this condition is not necessary to prove error bounds, but in particular it guarantees existence and uniqueness of the discrete solution of (5.5).

We remark here that an accurate nonconforming discretizations also requires a small *consistency error*. For some of the example element pairs below, this has been proven already, cf. [ANS01b] for example. Nevertheless this question is not of primary interest in our work.

Crouzeix-Raviart property for nonconforming approximation: Here we demand the so-called *Crouzeix-Raviart property* for $V_{\text{veloc}} \not\subset V$, i.e. the average jump of the discrete solution across a face vanishes:

(CR) $\displaystyle\int_E [\![\underline{v}_h]\!]_E = \underline{0} \qquad \forall \underline{v}_h \in V_{\text{veloc}}, \ \forall E \in \mathcal{E}.$

Note that for boundary faces E this simplifies to $\int_E \underline{v}_h = \underline{0}$.

5.3   Examples of finite element spaces

In this subsection we present a (nonexhaustive) list of finite element pairs fulfilling the theoretical assumptions (G1), (G2) and (CR) of the previous section. The exposition is kept to a minimum; details and proofs can be found in [CKN03]. First, the table below gives a brief overview of the element pairs.

Section	Type	spatial dim.	Elements	Remarks/Name
5.3.1	nonconform	2D+3D	triangles, tetrahedra	standard CR space
5.3.2	nonconform	2D	rectangles	enriched CR space
5.3.3	nonconform	2D	rectangles	enriched CR space
5.3.4	nonconform	3D	pentahedra	
5.3.5	nonconform	3D	hexahedra	
5.3.6	conform	2D	triangles, rectangles	Bernardi-Fortin-Raugel

As in the standard theory, a finite element is denoted by a triplet $(T, \mathcal{P}, \Sigma)$, where T is a domain, $\mathcal{P}$ denotes a space of functions, and Σ is a set of functionals of $\mathcal{P}^*$, cf. [Cia78].

5.3.1   Crouzeix-Raviart elements I

For a triangulation of Ω consisting of *triangles* in 2D or of *tetrahedra* in 3D, we approximate the velocity in the Crouzeix-Raviart finite element space and the pressure in the space of piecewise constant functions, namely (cf. [CR73, GR86, ANS01b])

$$V_{\text{veloc}} := \{\underline{v}_h \in [L^2(\Omega)]^d : \underline{v}_h|_T \in [\mathbb{P}^1]^d \ \forall T, \int_E [\![\underline{v}_h]\!]_E = \underline{0} \ \forall E\}, \tag{5.8}$$

$$Q_{\text{pre}} := \{q_h \in L_0^2(\Omega) : q_h|_T \in \mathbb{P}^0 \ \forall T\}. \tag{5.9}$$

Assumptions (G1), (G2) and (CR) are satisfied, cf. [CKN03].

5.3.2   Crouzeix-Raviart elements II

Here we restrict ourselves to the 2D case and to a triangulation of Ω made of *rectangles*. Due to the condition (G1) we actually need to modify the finite element given in [ANS01a, ANS01b]. On the reference rectangle $\bar{T} = (0,1)^2$ we define $\bar{\mathbb{Q}}^{1+} := \text{span } \{1, \bar{x}, \bar{y}, \bar{x}\bar{y}, \bar{y}^2\}$.

As degree of freedom (i.e. functionals of Σ) we take

$$\bar{l}_i(q) := \int_{\bar{E}_i} q, i = 1, \ldots, 4, \qquad \bar{l}_5(q) := \int_{\bar{T}} \bar{q}_5 q,$$

where $\bar{E}_i$ are the four edges of $\bar{T}$, and $\bar{q}_5(\bar{x}, \bar{y}) := 3(2\bar{x} - 1)(2\bar{y} - 1)$.

One readily checks that the triplet $(\bar{T}, \bar{\mathbb{Q}}^{1+}, \{\bar{l}_i\}_{i=1}^5)$ is a finite element. The finite element $(T, \mathbb{Q}^{1+}, \{l_i\}_{i=1}^5)$ on the actual anisotropic rectangle T is obtained by a standard affine transformation from $(\bar{T}, \bar{\mathbb{Q}}^{1+}, \{\bar{l}_i\}_{i=1}^5)$ such that $\bar{y}$ is mapped onto the stretching direction of the rectangle.[3]

Now we define the approximation velocity space by

$$V_{\text{veloc}} := \{\underline{v}_h \in [L^2(\Omega)]^2 : \underline{v}_h|_T \in [\mathbb{Q}^{1+}]^2 \ \forall T, \int_E [\![\underline{v}_h]\!]_E = \underline{0} \ \forall E\}, \tag{5.10}$$

and take Q_{pre} as above in (5.9).

Again assumptions (G1), (G2) and (CR) are satisfied, cf. [CKN03].

---

[3]Note that the degrees of freedom l_i of the actual element become *mean* integrals (over E or T).

5.3.3   Crouzeix-Raviart elements III

As above we consider the 2D case and a triangulation of Ω made of *rectangles*. For the previous element, the local velocity space $V_{\text{veloc}}|_T$ depends on the stretching direction of the rectangle T. Here we modify the element such that this dependence on the directionality is removed.

Consider the reference rectangle $\bar{T} = (0,1)^2$, set $\bar{\mathcal{P}} := \mathbb{P}^2$, and define the degrees of freedom (with the same notation as before) by

$$\bar{l}_i(q) := \int_{\bar{E}_i} q,\, i = 1, 2, 3, 4, \qquad \bar{l}_5(q) := \int_{\bar{T}} q\bar{q}_5, \qquad \bar{l}_6(q) := \int_{\bar{T}} q,$$

with $\bar{q}_5$ as above. One easily checks that the triplet $(\bar{T}, \bar{\mathcal{P}}, \{\bar{l}_i\}_{i=1}^6)$ is a finite element. On an anisotropic rectangle T we take the finite element $(T, \mathbb{P}^2, \{l_i\}_{i=1}^6)$ obtained by a standard affine transformation. The approximate velocity space is given by

$$V_{\text{veloc}} := \{\underline{v}_h \in [L^2(\Omega)]^2 : \underline{v}_h|_T \in [\mathbb{P}^2]^2 \,\forall T, \int_E [\![\underline{v}_h]\!]_E = \underline{0} \,\forall E\}, \tag{5.11}$$

and use Q_{pre} as defined by (5.9). The stability (G2) of this element pair has been shown in [CKN03]. Assumptions (G1) and (CR) hold as well.

5.3.4   Modified Crouzeix-Raviart elements on pentahedra

We restrict ourselves to the 3D case and to a triangulation of Ω made of rectangular *pentahedra*. We want to build a nonconforming approximation of V. Due to the condition (G1) we need to modify the finite element given in [ANS01a, ANS01b]. Indeed on a (reference) pentahedron the velocity space has to contain the space spanned by $1, \bar{x}, \bar{y}, \bar{z}, \bar{x}\bar{z}, \bar{y}\bar{z}$, which is not the case of the space introduced in [ANS01a, ANS01b]. In view of the condition (CR) in the reference pentahedron $\bar{T}$, we then need to construct a finite element whose set $\bar{\Sigma}$ of degrees of freedom contains the mean on the five faces.

Now we take $\bar{\mathcal{P}} := \mathbb{P}^2$ and $\bar{\Sigma} := \{\bar{l}_i\}_{i=1}^{10}$ defined by

$$\bar{l}_i(p) := \int_{\bar{E}_i} p, \text{ for } i = 1, \ldots, 5, \qquad \bar{l}_6(p) := \int_{\bar{T}} p, \qquad \bar{l}_{i+6}(p) := \int_{\bar{T}} p\tilde{q}_i, \text{ for } i = 1 \ldots 4,$$

where

$$\tilde{q}_1(\bar{x}, \bar{y}, \bar{z}) := 1 - 3\bar{x} - 2\bar{z} + 6\bar{x}\bar{z}, \qquad \tilde{q}_2(\bar{x}, \bar{y}, \bar{z}) := 1 - 3\bar{y} - 2\bar{z} + 6\bar{y}\bar{z},$$
$$\tilde{q}_3(\bar{x}, \bar{y}, \bar{z}) := 1 - 4\bar{x} - 2\bar{y} + 6\bar{x}\bar{y} + 3\bar{x}^2, \qquad \tilde{q}_4(\bar{x}, \bar{y}, \bar{z}) := 2\bar{x} - 2\bar{y} - 3\bar{x}^2 + 3\bar{y}^2.$$

As before one easily shows that the triplet $(\bar{T}, \mathbb{P}^2, \{\bar{l}_i\}_{i=1}^{10})$ is a finite element. Consider now the actual (anisotropic) pentahedron T which can be obtained from the reference pentahedron by an affine transformation. In this way also the finite element $(T, \mathcal{P}, \Sigma)$ is defined.

The approximation velocity space V_{veloc} becomes

$$V_{\text{veloc}} := \{\underline{v}_h \in [L^2(\Omega)]^3 : \underline{v}_h \in [\mathbb{P}^2]^3 \,\forall T, \int_E [\![\underline{v}_h]\!]_E = \underline{0} \,\forall E\}, \tag{5.12}$$

and define the approximation pressure space Q_{pre} again by (5.9). The stability (G2) of this element pair has been proven in [CKN03]. Assumptions (G1) and (CR) hold by definition.

5.3.5 Modified Crouzeix-Raviart elements on hexahedra

Here we consider the 3D case and a triangulation of Ω made of rectangular *hexahedra*. Inspired from the above subsection and the condition (G1), on the reference hexahedron $\bar{T} = (0,1)^3$ we take $\bar{\mathcal{P}} := \mathbb{P}^2 + \text{span}\{\bar{x}\bar{y}\bar{z}\}$ and $\bar{\Sigma} := \{\bar{l}_i\}_{i=1}^{11}$ defined by

$$\bar{l}_i(p) := \int_{\bar{E}_i} p, \text{ for } i = 1, \ldots, 6, \qquad \bar{l}_7(p) := \int_{\bar{T}} p, \qquad \bar{l}_{i+7}(p) := \int_{\bar{T}} p\tilde{q}_i, \text{ for } i = 1 \ldots 4,$$

where

$$\tilde{q}_1(\bar{x}, \bar{y}, \bar{z}) := 1 - 2\bar{x} - 2\bar{z} + 4\bar{x}\bar{z}, \qquad \tilde{q}_2(\bar{x}, \bar{y}, \bar{z}) := 1 - 2\bar{y} - 2\bar{z} + 4\bar{y}\bar{z},$$
$$\tilde{q}_3(\bar{x}, \bar{y}, \bar{z}) := 1 - 2\bar{x} - 2\bar{y} + 4\bar{x}\bar{y}, \qquad \tilde{q}_4(\bar{x}, \bar{y}, \bar{z}) := 1 - \bar{x} - \bar{y} - \bar{z} + 4\bar{x}\bar{y}\bar{z}.$$

As before the triplet $(\bar{T}, \bar{\mathcal{P}}, \{\bar{l}_i\}_{i=1}^{11})$ is a finite element. The finite element $(T, \mathcal{P}, \Sigma)$ on the actual hexahedron T is obtained by the usual affine transformation.

To define the pair $(V_{\text{veloc}}, Q_{\text{pre}})$, set[4]

$$V_{\text{veloc}} := \left\{ \underline{v}_h \in [L^2(\Omega)]^3 : \underline{v}_h|_T \in [\mathbb{P}^2 + \text{span}\{xyz\}]^3 \; \forall T, \int_E [\![\underline{v}_h]\!]_E = \underline{0} \; \forall E \right\}, \qquad (5.13)$$

and utilize Q_{pre} defined by (5.9). With a similar analysis as before, the stability (G2) of this element pair has been obtained without any assumption on the mesh, see [CKN03]. Furthermore this pair satisfies (G1) and (CR) by definition.

Now we present a family of *conforming* elements which satisfies the assumptions (G1) and (G2).

5.3.6 Bernardi-Fortin-Raugel elements

The Bernardi-Fortin-Raugel elements yield *conforming* discretizations. Here we restrict ourselves to the 2D case and to a triangulation of Ω made of *triangles* or *rectangles*. The discrete pressure space Q_{pre} is the space of piecewise constant functions defined by (5.9), and the approximate velocity space is defined by (cf. [GR86, AN])

$$V_{\text{veloc}} := \{\underline{v}_h \in [H_0^1(\Omega)]^2 : \underline{v}_h|_T \in \mathcal{P}_T \; \forall T\}. \qquad (5.14)$$

To define the local space $\mathcal{P}_T$ properly, start with triangular elements and consider an edge $E_i \subset \partial T$. Then let $\underline{p}_{E_i}$ be the edge bubble function 'in the direction of the normal vector' $\underline{n}_{E_i}$, i.e.

$$\underline{p}_{E_i} := \underline{n}_{E_i} b_{E_i,T}$$

(recall the definition of the edge bubble function $b_{E_i,T}$ from Chapter 2). The local space $\mathcal{P}_T$ then consists of linear functions enriched by the 'normal vector edge bubble functions' from above, namely

$$\mathcal{P}_T := [\mathbb{P}^1]^2 \oplus \text{span}\{\underline{p}_{E_i}\}_{i=1}^3.$$

For rectangular elements proceed similarly. Set again $\underline{p}_{E_i} := \underline{n}_{E_i} b_{E_i,T}$, where $b_{E_i,T}$ is now the edge bubble function for the rectangle T (see [CKN03, Section 4.1] for details). The local space $\mathcal{P}_T$ then becomes

$$\mathcal{P}_T := [\mathbb{Q}^1]^2 \oplus \text{span}\{\underline{p}_{E_i}\}_{i=1}^4.$$

For both elements, condition (G1) is fulfilled by definition. The stability of the Bernardi-Fortin-Raugel element is shown in [AN] for some families of meshes, i.e. assumption (G2) can be satisfied too.

---

[4]For the ease of notation, the definition of V_{veloc} is given for axiparallel meshes. Otherwise the span$\{xyz\}$ has to be replaced by span$\{\bar{x}\bar{y}\bar{z} \circ F_T^{-1}\}$, with F_T being the affine transformation from $\bar{T}$ to T.

5.4   Residual error estimators

The general philosophy of residual error estimators is to estimate an appropriate norm of the correct residual by terms that can be evaluated easier, and that involve the data at hand. To this end denote the *exact element residual* by

$$\underline{R}_T := \underline{f} - (-\Delta \underline{u}_h + \nabla p_h) \text{ on } T.$$

As it is common [Ver96], this exact residual is replaced by some finite dimensional approximation called *approximate element residual* $\underline{r}_T$,

$$\underline{r}_T \in \mathbb{P}^k(T) \text{ on } T.$$

Depending on the polynomial order k and the actual finite element, this approximation can be achieved by projecting either $\underline{f}$ alone or $\underline{R}_T$ as a whole.

Next, consider a face E (or edge in 2D) and recall that one normal unitary vector $\underline{n}_E$ is associated with it, cf. Chapter 2. Furthermore in the 2D case introduce temporarily the *tangent vector* $\underline{t}_E := \underline{n}_E^\perp$ that is orthogonal to $\underline{n}_E$. Now we are ready to introduce the gradient jump in normal and tangential direction by

$$\underline{J}_{E,n} := \begin{cases} [\![(\nabla \underline{u}_h - p_h I)\underline{n}_E]\!]_E & \text{for interior edges/faces} \\ \underline{0} & \text{for boundary edges/faces} \end{cases}$$

$$\underline{J}_{E,t} := [\![\nabla \underline{u}_h \, \underline{t}_E]\!]_E \qquad \text{for nonconforming 2D case (recall } \underline{t}_E = \underline{n}_E^\perp)$$
$$\underline{J}_{E,t} := [\![\nabla \underline{u}_h \times \underline{n}_E]\!]_E \qquad \text{for nonconforming 3D case,}$$

where the matrix-vector cross product $\nabla \underline{u}_h \times \underline{n}_E$ is defined in a rowwise fashion by

$$\nabla \underline{u}_h \times \underline{n}_E := \begin{bmatrix} (\nabla u_{h,1} \times \underline{n}_E)^\top \\ (\nabla u_{h,2} \times \underline{n}_E)^\top \\ (\nabla u_{h,3} \times \underline{n}_E)^\top \end{bmatrix} \in \mathbb{R}^{3\times 3}.$$

For nonconforming discretizations the tangential jump does not vanish on boundary faces. Following [Ver89, DDP95, KS00], abbreviate the error in the velocity and in the pressure by

$$\underline{e} := \underline{u} - \underline{u}_h, \qquad \varepsilon := p - p_h.$$

Now we are ready to define the error estimators.

Definition 5.4.1 (Residual error estimators)
For a conforming discretization, the local residual error estimators *is defined by*

$$\eta_T^2 := h_{min,T}^2 \|\underline{r}_T\|_T^2 + \|\text{div } \underline{u}_h\|_T^2 + \sum_{E \subset \partial T} \frac{h_{min,T}^2}{h_E} \|\underline{J}_{E,n}\|_E^2. \tag{5.15}$$

For a nonconforming 2D discretization we set

$$\eta_T^2 := h_{min,T}^2 \|\underline{r}_T\|_T^2 + \|\text{div } \underline{u}_h\|_T^2 + \sum_{E \subset \partial T} \frac{h_{min,T}^2}{h_E} \left(\|\underline{J}_{E,n}\|_E^2 + \|\underline{J}_{E,t}\|_E^2 \right), \tag{5.16}$$

while for a nonconforming 3D discretization the definition becomes

$$\eta_T^2 := h_{min,T}^2 \|\underline{r}_T\|_T^2 + \|\text{div } \underline{u}_h\|_T^2 \;+\; \sum_{E \subset \partial T} \frac{h_{min,T}^2}{h_E} \left(\|\underline{J}_{E,n}\|_E^2 + \|\underline{J}_{E,t}\|_E^2 \right) \tag{5.17}$$

$$+ \sum_{E \subset \partial T, E \in \mathcal{E}_\square} \frac{h_{min,T}^2}{h_E \, \varrho(E)^2} \|[\![\underline{u}_h]\!]_E\|_E^2,$$

where the last sum is over the rectangular faces alone.

The global residual error estimator *is given by*

$$\eta^2 := \sum_{T \in \mathcal{T}_h} \eta_T^2.$$

Furthermore denote the local and global approximation terms *by*

$$\zeta_T^2 := \sum_{T' \subset \omega_T} h_{min,T'}^2 \|\underline{r}_{T'} - \underline{R}_{T'}\|_{T'}^2, \qquad \zeta^2 := \sum_{T \in \mathcal{T}_h} \zeta_T^2.$$

5.4.1   Lower error bounds

The general framework to derive the *lower error bound* is to some extend standard (see [Ver96, DDP95] for the isotropic 2D counterpart). All 3D considerations seem to be novel. In particular nonconforming 3D discretizations with rectangular faces (i.e. the elements are pentahedra or hexahedra) display analytical technicalities which lead to (seemingly) suboptimal results. Recall now the notation for the velocity error $\underline{e} = \underline{u} - \underline{u}_h$ and the pressure error $\varepsilon = p - p_h$.

Theorem 5.4.2 (Local lower error bound) *Assume one of the following cases: The discretization is 2D, or the discretization is 3D and conforming, or the discretization is 3D, is nonconforming and is only composed of tetrahedra. Then for all elements T, the following local lower error bound holds:*

$$\eta_T \lesssim \|\nabla_T \underline{e}\|_{\omega_T} + \|\varepsilon\|_{\omega_T} + \zeta_T. \tag{5.18}$$

Key ideas of the proof: The proof is given in [CKN03]. It relies on standard techniques (e.g. bubble functions, partial integration, and inverse inequalities) that are adapted to accommodate the anisotropy where necessary.   ∎

The remaining case covers nonconforming 3D discretizations with rectangular faces. Then the error estimator (5.17) contains the jump term $[\![\underline{u}_h]\!]_E$ of the discrete solution across a rectangular face. Unfortunately we failed to bound this term locally; only a weaker global bound could be established. This is mainly caused by the anisotropy of the elements, i.e. for isotropic discretizations this special case does not occur.

Theorem 5.4.3 (Global lower error bound) *For a 3D nonconforming triangulation consisting of pentahedra or hexahedra, the following global lower error bound holds:*

$$\eta \lesssim m_1(\underline{e}, \mathcal{T}_h) \|\nabla_T \underline{e}\| + \|\varepsilon\| + \zeta.$$

Key ideas of the proof: The proof can be found in [CKN03]. It utilizes the Crouzeix-Raviart property (CR), a Poincaré like inequality, and a trace inequality.   ∎

5.4.2   Upper error bounds

For the nonconforming 2D case we proceed similar to [DDP95], with the necessary adaptations due to the anisotropy of our discretization. The whole 3D analysis seems to be new. Basic steps are always partial integration, combined with Galerkin like orthogonalities and interpolation error estimates.

First we bound the pressure error (for conforming and nonconforming discretizations). The bound of the velocity error is divided in two parts since conforming and nonconforming discretizations are treated differently.

Lemma 5.4.4 (Error in the pressure) *There exists a function* $\underline{v}_\varepsilon \in [H_0^1(\Omega)]^d$ *depending on* $\varepsilon = p - p_h$ *such that the error in the pressure is bounded by*

$$\|\varepsilon\| \lesssim m_1(\underline{v}_\varepsilon, \mathcal{T}_h)(\eta + \zeta) + \|\nabla_\mathcal{T}\,\underline{e}\|. \tag{5.19}$$

Key ideas of the proof: For details of the proof we refer to [CKN03]. Starting point is a continuous inf-sup (like) condition applied to the pressure error ε. The Galerkin orthogonality, partial integration, and interpolation results complement the proof. ∎

Theorem 5.4.5 (Upper error bound - Conforming case.) *Assume a conform discretization, and let* $\underline{v}_\varepsilon \in [H_0^1(\Omega)]^d$ *be the function from Lemma 5.4.4. Then the error is bounded globally from above by*

$$\|\varepsilon\| + \|\nabla_\mathcal{T}\,\underline{e}\| \lesssim \Big(m_1(\underline{e}, \mathcal{T}_h) + m_1(\underline{v}_\varepsilon, \mathcal{T}_h)\Big)(\eta + \zeta). \tag{5.20}$$

Key ideas of the proof: The proof is contained in [CKN03]. Main ingredients are the previous lemma and again the Galerkin orthogonality, partial integration, and interpolation results. ∎

Theorem 5.4.6 (Upper error bound - Nonconforming case.) *Let* $\underline{v}_\varepsilon$ *be the function from Lemma 5.4.4. Furthermore there exist functions* $\underline{r}_\varepsilon \in [H_0^1(\Omega)]^2$, $\underline{s}_\varepsilon \in [H^1(\Omega)]^2$ *both depending on* $\underline{e} = \underline{u} - \underline{u}_h$ *such that the error is bounded globally from above by*

$$\|\varepsilon\| + \|\nabla_\mathcal{T}\,\underline{e}\| \lesssim \Big(m_1(\underline{r}_\varepsilon, \mathcal{T}_h) + m_1(\underline{s}_\varepsilon, \mathcal{T}_h) + m_1(\underline{v}_\varepsilon, \mathcal{T}_h)\Big)(\eta + \zeta). \tag{5.21}$$

In the 3D case the vector function $\underline{s}_\varepsilon$ *is replaced by a matrix function* $\underline{\underline{s}}_\varepsilon \in [H^1(\Omega)]^{3\times 3}$.

Key ideas of the proof: For the proof see [CKN03].

Due to the nonconforming discretization, a Helmholtz like decomposition of the velocity error $\underline{e}$ plays a vital role and eventually introduces the functions $\underline{r}_\varepsilon$ and $\underline{s}_\varepsilon$. As usual the Galerkin orthogonality, partial integration, and interpolation results are utilized.

In contrast to conforming discretizations, one also employs a particular error orthogonality (with the curl of some interpolant). In 3D this leads to the jump $[\![\underline{u}_h]\!]_E$ across rectangular faces, cf. (5.17). ∎

Remark 5.4.7 (Alignment measure) The upper error bounds (5.20) and (5.21) contain several *alignment measures* $m_1(\cdot, \mathcal{T}_h)$. This is in contrast to estimators for isotropic meshes: For anisotropic discretizations, all known estimators are (explicitly or implicitly) based on an anisotropic mesh that is suitably aligned with the anisotropic function.

Compared with the isotropic estimators, our upper error bounds are special in the sense that the alignment measure cannot be evaluated explicitly. However, this should not be considered too much as a disadvantage. For example, the alignment measure $m_1(\underline{e}, \mathcal{T}_h)$ for the error $\underline{e} = \underline{u} - \underline{u}_h$ is of size $\mathcal{O}(1)$ for sufficiently good meshes, cf. [Kun00, Kun01e, Kun02b] and Chapter 6. We expect a similar behaviour for the other alignment measures. This confidence is strengthened by the numerical experiments below.

In practical computations one may simply use the error estimator without considering the alignment measures. For adaptive algorithms this is well justified since a local lower error bound holds unconditionally (with the exception of nonconforming 3D triangulations made of pentahedra or hexahedra), i.e. the estimator is efficient. □

Remark 5.4.8 (Isotropic discretizations) Our estimators can of course also be applied to *isotropic* discretizations. Then some definitions (and the corresponding analysis) simplify. Some known results are reproduced. Other estimators for certain element pairs seem to be novel. For a detailed description see [CKN03, Section 6.4]. □

5.5 Numerical experiments

The following experiments are taken from [CKN03] and have been carried out by E. Creusé. They will underline and confirm our theoretical predictions.

This example consists in solving the two dimensional Stokes problem (5.1) on the unit square $\Omega = (0, 1)^2$. Here, we use the Crouzeix-Raviart element II (see Section 5.3.2), on an anisotropic Shishkin type mesh composed of rectangles. This mesh is the tensor product of a 1D Shishkin type mesh and a uniform mesh, both with n subintervals. With $\tau \in (0, 1)$ being a transition point parameter, the coordinates (x_i, y_j) of the nodes of the rectangles are defined by

$$dx_1 := 2\tau/n, \qquad dx_2 := 2(1-\tau)/n, \qquad dy = 1/n,$$

$$\begin{cases} x_i := i\,dx_1 & (0 \le i \le n/2), \\ x_i := \tau + (i - n/2)\,dx_2 & (n/2 + 1 \le i \le n), \\ y_j := j\,dy & (0 \le j \le n). \end{cases}$$

τ

Figure 5.5.1: Shishkin type mesh on the unit square with $n = 8$.

The discrete problem (5.5) is solved with the Uzawa algorithm. The number of degrees of freedom is equal to $n(3n + 2)$ for each component of the velocity, and equals n^2 for the pressure. The total number of degrees of freedom (DoF) is then equal to $n(7n + 4)$. Here we utilize a sequence of meshes $\{\mathcal{T}_k\}_{k=1}^{8}$, where we set $n := 2^{k+1}$.

The tests are performed with the following prescribed exact solution $(\underline{u}, p)$ which depends on a parameter ϵ:

$$\begin{cases} \Phi &= x^2(1-x)^2 y^2(1-y)^2 e^{-x/\sqrt{\epsilon}}, \\ \underline{u} &= \underline{\mathrm{curl}}\ \Phi, \\ p &= e^{-x/\sqrt{\epsilon}}. \end{cases}$$

This allows to have in particular $\mathrm{div}\,\underline{u} = 0$ and $\underline{u}_{|\Gamma} = 0$. Note that $\underline{u}$ and p present an exponential boundary layer of width $\mathcal{O}(\sqrt{\epsilon}|\ln\sqrt{\epsilon}|)$ along the line $x = 0$. The transition parameter τ involved in the construction of the Shishkin-type mesh is defined by $\tau := \min\{1/2, 2\sqrt{\epsilon}|\ln\sqrt{\epsilon}|\}$, i.e. it is roughly twice the boundary layer width. The maximum aspect ratio in the mesh is equal to $1/(2\tau)$.

To begin with, let us check that the numerical solution $(\underline{u}_h, p_h)$ converges towards the exact one. To this end we plot the curves

- $\|\nabla_\mathcal{T}(\underline{u} - \underline{u}_h)\|_\Omega$ as a function of DoF (see Figure 5.5.2 left),
- $\|p - p_h\|_\Omega$ as a function of DoF (see Figure 5.5.2 right).

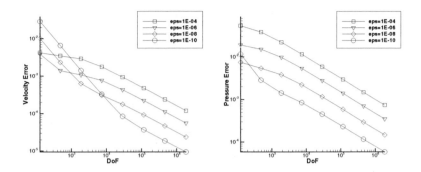

Figure 5.5.2: $\|\nabla_h(\underline{u} - \underline{u}_h)\|_\Omega$ (left) and $\|p - p_h\|_\Omega$ (right) in dependence of DoF.

As we can see, the convergence rates for the velocity and for the pressure are of order 0.5, as theoretically expected. This shows the good convergence of $(\underline{u}_h, p_h)$ towards $(\underline{u}, p)$.

Now we investigate the main theoretical results which are the upper and the lower error bounds. In order to present the underlying inequalities (5.18) and (5.21) appropriately, we reformulate them by defining the ratios of left-hand side and right-hand side, respectively:

- $q_{up} = \dfrac{\|\nabla_\mathcal{T}(\underline{u} - \underline{u}_h)\|_\Omega + \|p - p_h\|_\Omega}{\eta + \zeta}$ as a function of DoF,

- $q_{low} = \max\limits_{T \in \mathcal{T}_h} \dfrac{\eta_T}{\|\nabla_\mathcal{T}(\underline{u} - \underline{u}_h)\|_{\omega_T} + \|p - p_h\|_{\omega_T} + \zeta_T}$ as a function of DoF.

The second ratio is related to the local lower error bound and measures the *efficiency* of the estimator. According to Theorem 5.4.2, q_{low} has to be bounded from above. This can be observed indeed in the right part of Figure 5.5.3. Hence the estimator is *efficient*. Note that the values of q_{low} are much alike the ones for other problem classes, cf. [Kun00, Kun01e].

The first ratio q_{up} is frequently referred to as *effectivity index*. It measures the *reliability* of the estimator and is related to the global upper error bound. In order to investigate this error bound, recall first that the alignment measures $m_1(\cdot, \mathcal{T}_h)$ are expected to be of moderate size since we employ well adapted meshes (cf. Theorem 5.4.6). Hence the corresponding ratio q_{up} should be bounded from above which is confirmed by the experiment (left part of Figure 5.5.3). As soon as a reasonable resolution of the layer is achieved, the quality of the upper error bound is independent of ϵ. Thus the estimator is *reliable*. Again the values of q_{up} resemble the ones for other problem classes, cf. above.

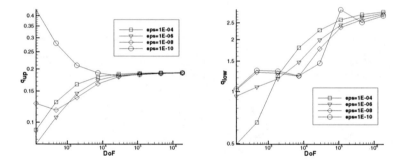

Figure 5.5.3: q_{up} (left) and q_{low} (right) in dependence of DoF.

Chapter 6

Anisotropic mesh versus anisotropic function

The material of this chapter originates from [Kun02b]. It investigates the relation between anisotropic error estimation and anisotropic meshes and their generation. The alignment measure is of particular interest.

The relation to other chapters is clearly given by the anisotropic solution, the anisotropic discretization, and the interplay with error estimation.

6.1 Introduction

Imagine some problem (PDE) with an anisotropic exact solution u. Then one seeks an appropriate anisotropic discretization $\mathcal{T}_h$. Starting from this setting, we will approach the topic of this chapter from two sides. Firstly we consider anisotropic error estimation and the alignment measure in particular (cf. the previous chapters). For the second approach we scrutinize anisotropic adaptive algorithms.

Approach via error estimation

Let us recall the main results of the previous chapters. In Theorems 3.3.4, 3.4.4, 3.5.2, 3.5.5, 4.2.4, 4.3.4 and partially in Theorem 5.4.5 , upper and lower error bounds are presented. The upper error bound always contains the *alignment measure* $m_1(u-u_h, \mathcal{T}_h)$ that measures the suitability of the anisotropic discretization $\mathcal{T}_h$.

The main problem in anisotropic error estimation is to derive upper and lower error bounds that show asymptotically equivalent behaviour. In other words, both error bounds should be as tight as possible. For the upper error bound this demand requires a small alignment measure $m_1(u-u_h, \mathcal{T}_h) \sim 1$. In numerous practical examples we obtained values in the range $2 \ldots 4$ if the anisotropic mesh is sufficiently well aligned with the anisotropy of the solution (in a rather weak sense), see also Section 3.6.

This leads directly to a main question: Is it possible to *construct* an anisotropic mesh that *guarantees* a small alignment measure $m_1(u - u_h, \mathcal{T}_h) \sim 1$? With some heuristic assumptions, the answer will be 'Yes'.

Approach via anisotropic adaptive algorithm

For our second kind of motivation we scrutinize anisotropic adaptive algorithms. For clarity let us recall a standard isotropic version first.

74

Isotropic adaptive algorithm:

0. Start with an initial mesh $\mathcal{T}_0$.
1. Solve the corresponding discrete system.
2. Compute the local *a posteriori* error estimator for each element T of the mesh.
3. When the estimated global error is small enough then stop.
   Otherwise obtain information for a new, better mesh, namely the element size (as a function over Ω).
4. Based on this information, construct a new mesh or perform a mesh refinement, and re-iterate.

Assume next an anisotropic solution. Hence an anisotropic discretization is sought, and the adaptive procedure has to be modified as follows (cf. also the introduction of Chapter 1).

Anisotropic adaptive algorithm:

0. Start with an initial mesh $\mathcal{T}_0$.
1. Solve the corresponding discrete system.
2. Compute the local *a posteriori* error estimator for each element T of the mesh.
3. When the estimated global error is small enough then stop.
   Otherwise obtain information for a new, better mesh. This includes:
   - Detect regions of anisotropic behaviour of the solution.
   - Determine a (quasi) optimal stretching direction and stretching ratio of the finite elements in those regions.
   - Determine the element size.
4. Based on this information, construct a new mesh or perform a mesh refinement, and re-iterate.

Implementational and analytical aspects of anisotropic solutions and meshes are given in more detail e.g. in [Ape99, BK94, Dol98, FLR96, FPZ01, FG00, HDB⁺00, KR90, Kra01, Noc95, PVMZ87, Rac93, RGK93, Sie96, Sim94, VH96, ZW94] or in Kunert [Kun99] and the literature cited therein.

Here it shall suffice to mention that all steps of the adaptive algorithms have to be reinvestigated. The efficient solution of large systems (step 1) requires advanced methods and/or preconditioners (such as multi-grid or multi-level methods; and in particular for 3D domains). The difficulties with anisotropic error estimators (step 2) have been addressed in the previous chapters. The necessary changes with mesh generation (step 4) are readily visible, since now anisotropic elements (with prescribed stretching) have to be constructed.

Equally obvious are the modifications required for the information extraction of step 3. This will be the main focus from now on. Ideally, the required anisotropic information should be provided by an error estimator (which has been investigated thoroughly). Reality, however, is different. Unfortunately almost all anisotropic information can not be derived from the error estimator but are provided by external procedures instead. The following scheme depicts this.

Source:	Error estimator	External information extraction (e.g. Hessian strategy; Contour lines; ...)
	$\Downarrow$	$\Downarrow$
Obtained anisotropic information:	• element size	• stretching direction • stretching ratio • (element size)

There are several ways to obtain the required anisotropic information. The most popular approaches are:

- **Hessian strategy:** Approximate the Hessian $D^2 u$ and perform a *spectral analysis* (also known as *principal axes transformation*). The eigenvectors tell the stretching directions; the eigenvalues give the aspect ratio. See [CHM95, Dol98, PVMZ87, RGK93, Sim94, ZW94] for a more detailed description.

  In [FP00] a similar approach is pursued. With some heuristic assumptions, an upper error bound is proven but not a lower error bound.

- **Level lines:** The *level lines* (or *contour lines*) provide a vivid picture of the anisotropy of the function, and thus of the stretching direction. The numerical realization goes back to Kornhuber and Roitzsch [KR90].

- **Gradient jump:** The *gradient jump* of certain values give some indication in which direction the elements should be stretched, see [BK94, Rac93, Sie96].

  A different methodology is used in [AGJM01] where the error contributions in different directions are balanced. These contributions are approximated by local problems. First experiments gave promising results.

In this chapter we will solely concentrate on the Hessian strategy. The main question is now: When the anisotropic mesh is constructed via the Hessian strategy, will this new mesh be suitable for error estimation (in the sense of $m_1(u - u_h, \mathcal{T}_h) \sim 1$)?

Both approaches to the topic of this work lead to similar questions. The common answer and the aim of this work can be formulated as follows.

> *With only few heuristic assumptions we show that an anisotropic mesh $\mathcal{T}_{\text{new}}$ constructed via the Hessian strategy implies a small alignment measure, i.e. $m_1(u - u_h, \mathcal{T}_{\text{new}}) \lesssim 1$.*

Then one can conclude that reliable error estimation is possible, cf. Theorem 3.3.4 and the other error estimation theorems mentioned above. The value of our work is twofold.

- The unknown alignment measure $m_1(u - u_h, \mathcal{T}_h)$ of the upper error bound is not as disadvantageous as it may seem at first sight. In contrast, suitable mesh construction techniques (e.g. based on the Hessian strategy) automatically imply a small alignment measure $m_1(u - u_h, \mathcal{T}_{\text{new}}) \lesssim 1$.
- The heuristic Hessian strategy is based upon interpolation error considerations. We provide theoretical justification that this strategy also provides a small alignment measure. Hence the meshes allow reliable error estimation.

We may add that our contribution is primarily of theoretical interest. The main practical implication is that now we can apply all the aforementioned error estimators without having to worry (too much) about the unknown size of the alignment measure. With some heuristic assumptions, known anisotropic mesh construction techniques will provide discretizations that allow sharp error bounds.

The remainder of this chapter is organised as follows. Some additional notation is introduced in Section 6.2. In Section 6.3 the Hessian strategy is motivated and described. Finally Section 6.4 is devoted to the heuristic assumptions. The main result is presented there, and we comment on the numerical experiences.

6.2 Notation

In this chapter we consider exclusively tetrahedral (3D) or triangular (2D) elements. The notation of the elements is as described in Chapter 2, cf. also Figure 2.1.1.

In the analysis, derivatives of certain functions play an important role. The so–called *Hessian* is the matrix of the second–order derivatives, denoted by

$$D^2 u := \left(\partial_{x_i} \partial_{x_j} u \right)_{i,j=1}^{d}$$

where $\partial_{x_i} v$ is the partial first order derivative.

When analysing the anisotropic error estimates, certain *directional derivatives* can be favourably employed, cf. [Kun99]. For arbitrary directions (vectors) l_1, l_2 we thus define the first order directional derivative by

$$\partial_{l_1} u := |l_1|^{-1} \cdot l_1^{\top} \nabla u \equiv |l_1|^{-1} \cdot (\nabla u, l_1) \equiv \partial u / \partial l_1 \quad ,$$

and the second order directional derivative by

$$\partial_{l_1 l_2}^2 u := |l_1|^{-1} |l_2|^{-1} \cdot l_1^{\top} D^2 u \, l_2 \equiv |l_1|^{-1} |l_2|^{-1} \cdot (D^2 u \, l_1, l_2) \equiv \partial^2 u / \partial l_1 \partial l_2 \quad ,$$

where $|l_i|$ denotes the length of the vector and $(\cdot, \cdot)$ is the usual Euclidean scalar product. Note that the derivatives are with respect to the *unitary* direction, even if the direction vector l_i has arbitrary length.

In the Hessian strategy which is described in the next section the Hessian $D^2 u$ naturally is of great importance. However in real applications $D^2 u$ is not known since it involves the exact solution u. Hence a so–called *approximate Hessian*

$$D^{2,h} u_h \approx D^2 u$$

has to be computed from the known approximate solution u_h. This can be achieved via a recovered gradient, for example. Here we do not discuss this question but assume instead that a symmetric and sufficiently good approximation $D^{2,h} u_h$ is provided (see e.g. [AV02] for more details). In Section 6.4 we explain what 'sufficiently good' means.

For the approximate Hessian we want to utilize the equivalent of the directional derivatives and therefore define

$$\partial_{l_1 l_2}^{2,h} u_h := |l_1|^{-1} |l_2|^{-1} \cdot l_1^{\top} D^{2,h} u_h \, l_2 \equiv |l_1|^{-1} |l_2|^{-1} \cdot (D^{2,h} u_h \, l_1, l_2) \quad .$$

Later on the *Crouzeix–Raviart interpolation* will be required. The interpolation operator $I_{CR} : H^1(\Omega) \mapsto L^2(\Omega)$ is uniquely defined by the condition

$$(I_{CR} v)|_T \in \mathbb{P}^1(T) \qquad \text{and} \qquad \int_E v - I_{CR} v = 0 \qquad \forall \text{ faces } E \subset \partial T \quad .$$

An anisotropic interpolation error bound that has been proven in [ANS01a, Lemma 3.3] reads as follows.

Lemma 6.2.1 *Let $u \in H^2(T)$. Then*

$$\| \partial_{p_i} (u - I_{CR} u) \|_T \lesssim \sum_{j=1}^{3} h_{j,T} \cdot \| \partial_{p_i p_j}^2 u \|_T \qquad i = 1, 2, 3. \tag{6.1}$$

Note that the additional assumptions of [ANS01a, Lemma 3.3] are automatically satisfied here since we utilize the directional derivatives. Furthermore no maximum angle condition is required.

In order to have a unified notation in all chapters here, we have changed the notation of [Kun02b] slightly. Most notably the terms $H_i(x)$ and $\mathbf{q}_i(x)$ of [Kun02b] are here denoted by $h_i^{\text{pred}}(x)$ and $\mathbf{p}_i^{\text{pred}}(x)$, respectively. Furthermore in the shortened exposition of this chapter we do not require $\mathbf{p}_i^{\text{pred}}$ and $\mathbf{p}_{i,T}$ to be of unitary length, as it is the case in [Kun02b].

6.3   Hessian strategy

The Hessian strategy has been known for quite a long time, see e.g. [CHM95, PVMZ87, Sim94, ZW94]. For a given anisotropic function it provides the description of a suitable anisotropic mesh (in terms of stretching direction and ratio of the elements). A simple, heuristically convincing and straight–forward motivation is as follows.

Motivation (2D case): Consider a single triangle T and an arbitrary quadratic function v. Minimize the interpolation error $\|v - \mathrm{I}_{\text{Lagr}}v\|_*$ over all (right–angled) triangles T with prescribed area $|T|$. Here $\mathrm{I}_{\text{Lagr}}v \in \mathbb{P}^1(T)$ shall be the usual nodal Lagrange interpolate, and $\|\cdot\|_*$ is a norm to be specified. Compute the (constant) Hessian D^2v, its two real eigenvalues λ_1, λ_2 such that $|\lambda_1| \le |\lambda_2|$, and the corresponding eigenvectors $\mathbf{p}_1$ and $\mathbf{p}_2$. Then the quasi minimal interpolation error (up to some constant factor) is achieved for those triangles that are stretched along $\mathbf{p}_1$.

The (quasi) optimal stretching ratio depends on the choice of the norm $\|\cdot\|_*$. For the L^2 norm or the L_∞ norm the optimal ratio is $|\lambda_1|^{-1/2} : |\lambda_2|^{-1/2}$. This ratio is used by [CHM95, PVMZ87, Sim94], for example, and will be considered in this chapter too.

For a minimal interpolation error in the H^1 seminorm, however, the (quasi) optimal stretching ratio is $|\lambda_1|^{-1} : |\lambda_2|^{-1}$. This ratio does not seem to be used before, and it is not investigated here.

To describe the Hessian strategy properly, we formally split it into step 3 of the adaptive algorithm (information extraction) and step 4 (remeshing). This separation allows to address the assumptions appropriately but also the drawbacks. Our main interest is the information extraction (step 3). Furthermore several remeshing algorithms (step 4) are presented but not discussed in detail. Now the Hessian strategy is presented.

Step 3: Information extraction
Although the finite element error $\|\nabla(u - u_h)\|$ should be minimized, one considers instead the simpler interpolation error $\|\nabla(u - \mathrm{I}_{\text{Lagr}}u)\|$ which leads to the Hessian D^2u. Since D^2u is not known one has to utilize a symmetric approximation $D^{2,h}u_h$ to D^2u. We assume that such an approximation is provided.

The information desired are the stretching directions and the aspect ratio. Generally speaking, they are functions of $x \in \Omega$, and they are given by the eigenvalues and eigenvectors of $D^{2,h}u_h$. This procedure is known as *spectral decomposition* or *principal axes transformation* of the matrix $D^{2,h}u_h$. One computes

Eigenvectors of $D^{2,h}u_h$:	$\mathbf{p}_i^{\text{pred}}(x)$	$i = 1, 2, 3$	(6.2)

Eigenvalues of $D^{2,h}u_h$:
$$\mu_i(x) := \lambda_i(D^{2,h}u_h(x)) \qquad i = 1, 2, 3 \qquad (6.3)$$
$$\equiv \partial^{2,h}_{\mathbf{p}_i^{\text{pred}}\mathbf{p}_i^{\text{pred}}}u_h(x)$$
Assume $|\mu_1(x)| \leq |\mu_2(x)| \leq |\mu_3(x)|$.

For each point $x \in \Omega$ the information desired is:

Stretching directions:	$\mathbf{p}_i^{\text{pred}}(x)$	$i = 1, 2, 3$	(6.4)
Stretching lengths:	$h_i^{\text{pred}}(x) := \alpha_T \cdot |\mu_i(x)|^{-1/2}$	$i = 1, 2, 3$	(6.5)

Note that the eigenvalues $\mu_i(x)$ determine only the stretching *ratio*. The actual stretching *lengths* and hence the desired element size are determined by the local parameter α_T. It can be chosen e.g. from the error (or error estimator) on the present element, and the desired (or predicted) error on the new element.

For the ease of our exposition we additionally assume that all $\mu_i(x)$ are distinct, that is $|\mu_1(x)| < |\mu_2(x)| < |\mu_3(x)|$. This is, of course, not realistic in real world applications but much more convenient to describe our key ideas. If there were points $x^* \in \Omega$ with $\mu_1(x^*) = \mu_2(x^*)$ for example then one has a subspace of eigenvectors. The eigenvector $\mathbf{p}_1^{\text{pred}}(x)$, which is by definition related to the smallest absolute eigenvalue is thus likely to jump in the neighbourhood of that x^*. Then one would require a more sophisticated implementation and a different notation.

Step 4: Remeshing
Once the stretching directions and stretching lengths are known (as functions over Ω), the new mesh $\mathcal{T}_{\text{new}}$ is to be constructed. We will mention several approaches for the remeshing. The advantage of our exposition is that all of them can be treated (potentially) within the framework of our analysis. This is achieved by formulating a general condition that has be be satisfied. We note that this condition comprises most of the heuristic assumptions (or, in some sense, the essence of the Hessian strategy), cf. the discussion below.

When a new triangle $T_{\text{new}} \in \mathcal{T}_{\text{new}}$ is to be constructed[1], its stretching directions $\mathbf{p}_{i,T_{\text{new}}}$ and lengths $h_{i,T_{\text{new}}}$ should be close to the predicted stretching directions and lengths:[2]

$$\mathbf{p}_{i,T_{\text{new}}} \approx \mathbf{p}_i^{\text{pred}}(x) \qquad \forall x \in T_{\text{new}}$$
$$h_{i,T_{\text{new}}} \approx h_i^{\text{pred}}(x) \qquad \forall x \in T_{\text{new}} \quad .$$

The 'closeness' could be measured, for example, in some integral sense. It also displays an immediate difficulty: T_{new} should be constructed according to $\mathbf{p}_i^{\text{pred}}(x), h_i^{\text{pred}}(x)$, but these data should be considered for all points $x \in T_{\text{new}}$. This element T_{new}, however, is to be constructed and not known yet. This dependence (depicted by $\longleftarrow$) can be illustrated as follows:

$$T_{\text{new}} \quad \longleftarrow \quad \mathbf{p}_{i,T_{\text{new}}}, h_{i,T_{\text{new}}} \quad \longleftarrow \quad \mathbf{p}_i^{\text{pred}}(x)\big|_{T_{\text{new}}}, h_i^{\text{pred}}(x)\big|_{T_{\text{new}}} \quad \longleftarrow \quad T_{\text{new}}.$$

This self–dependent (reflexive) description can be dealt with in several ways.

---

[1] For easier recollection we denote the mesh to be constructed by $\mathcal{T}_{\text{new}}$.
[2] The 'closeness' of the vectors $\mathbf{p}_{i,T_{\text{new}}}$ and $\mathbf{p}_i^{\text{pred}}(x)$ concerns only their *direction* but not their length.

Firstly, one may iteratively construct the new element: Start with an initial guess $T_{new,1}$, determine $\mathbf{p}_i^{\mathrm{pred}}(x)\big|_{T_{new,1}}, h_i^{\mathrm{pred}}(x)\big|_{T_{new,1}}$, recompute $\mathbf{p}_{i,T_{new,2}}, h_{i,T_{new,2}}$, and construct an improved guess $T_{new,2}$ until a sufficient correspondence is achieved.

A second approach consists of constructing a global transformation such that the transformed domain can be meshed uniformly and isotropically, cf. [Sim94]. However such a transformation does not necessarily exist. Even if it exists, its construction requires solving a system of ordinary differential equations.

A third group of approaches heavily relies on the use of a Riemannian metric tensor; they try to construct equilateral triangles with respect to that metric. For more details or different methods see e.g. [BH96] or [Kun99] and the citations therein.

As already mentioned we will not present and analyse these remeshing algorithms in detail. Instead we want our analysis to include all these constructions, and probably even completely different approaches (e.g. the contour lines approach, cf. Section 6.1). For that reason we pose one rather general assumption: We allow *any* construction of $\mathcal{T}_{new}$ if for each element T_{new} its three stretching lengths and pure second order derivatives are balanced. This will become assumption (A.3') below:

$$h_{i,T_{new}}^2 \cdot |T_{new}|^{-1/2} \cdot \|\partial_{\mathbf{p}_i\mathbf{p}_i}^2 u\|_{T_{new}} \ \sim \ \alpha_T^2 \qquad i = 1, 2, 3, \tag{6.6}$$

or, in a weaker form, assumption (A.3) of Section 6.4. This condition contains much of the essence of the remeshing part of the Hessian strategy and its heuristic assumptions.

6.4   On the boundedness of the Alignment Measure

All adaptive algorithms with the Hessian strategy contain certain heuristic assumptions (probably in some disguise) which can be classified roughly as follows.

- The Hessian strategy is feasible,
- sufficiently good approximation of the Hessian: $D^{2,h}u_h \approx D^2 u$,
- the Hessian does not change rapidly across adjacent elements,
- the interpolation estimates are sharp enough,
- $u_h \approx I_{\mathrm{CR}}u$ in the sense of the alignment measure m_1.

Here we describe these assumptions verbally and in a mathematical style. More details are given in[Kun02b] where we also investigate the plausibility of the assumptions, describe potential difficulties, and explore the numerical behaviour of the assumptions.

Before starting, it is important to realize that the old mesh $\mathcal{T}_{\mathrm{old}}$ and the newly constructed mesh $\mathcal{T}_{\mathrm{new}}$ play a completely different role, as far as the heuristic assumptions are concerned:

- The old mesh $\mathcal{T}_{\mathrm{old}}$ is the basis for the Hessian strategy and gives the predicted stretching directions $\mathbf{p}_i^{\mathrm{pred}}(x)$ and stretching lengths $h_i^{\mathrm{pred}}(x)$.
  Using this information, the new mesh $\mathcal{T}_{\mathrm{new}}$ will be constructed.
- The new mesh $\mathcal{T}_{\mathrm{new}}$ has to satisfy the heuristic assumptions. In other words, all assumptions involving an element T are stated for elements $T \in \mathcal{T}_{\mathrm{new}}$.

These heuristic conditions are now explained and reformulated in a strict mathematical form.

1. **The Hessian strategy is feasible.**
   This implies that the Hessian can be computed, i.e.

   $$u \in H^2(\Omega) \qquad . \tag{A.1}$$

   Furthermore the stretching lengths are defined via $h_i^{\text{pred}}(x) := \alpha_T \cdot |\mu_i(x)|^{-1/2}$, cf. (6.5). Strictly speaking, this requires $\mu_i(x) \neq 0$. Here, however, we formally allow $\mu_i(x) = 0$ implying $h_i^{\text{pred}}(x) = \infty$. Then the remeshing part of the Hessian strategy has to deal with this 'exception'. The other assumptions (A.2)–(A.5) below have to hold as well, even if the heuristic reasoning may be less convincing (cf. assumption (A.3) for example).

2. **$D^{2,h}u_h \approx D^2u$ and D^2u does not change too much.**
   Assume for the moment that the Hessian D^2u is constant (i.e. u is quadratic), and that the directions $\mathbf{p}_{i,T_{new}}$ of the new element are chosen to be the eigenvectors of the exact Hessian D^2u (instead of its approximation $D^{2,h}u_h$). The principle of the spectral decomposition of the Hessian D^2u readily implies

   $$\partial^2_{\mathbf{p}_i\mathbf{p}_j}u = 0 \qquad \forall\, i \neq j \qquad .$$

   In reality, however, D^2u is rarely constant, and the directions $\mathbf{p}_{i,T_{new}}$ are computed from $D^{2,h}u_h$. Yet if D^2u does not change too much and if $D^{2,h}u_h \approx D^2u$ then $\partial^2_{\mathbf{p}_i\mathbf{p}_j}u$ should almost vanish. The amount which $\partial^2_{\mathbf{p}_i\mathbf{p}_j}u$ may deviate from zero is expressed relative to the lengths $h_{i,T}$ and the pure second order derivatives $\partial^2_{\mathbf{p}_i\mathbf{p}_i}u$ by the condition

   $$\sum_{i,j=1}^{3} h_{i,T}^2 \cdot h_{j,T}^2 \cdot \|\partial^2_{\mathbf{p}_i\mathbf{p}_j}u\|_T^2 \lesssim \sum_{i=1}^{3} h_{i,T}^4 \cdot \|\partial^2_{\mathbf{p}_i\mathbf{p}_i}u\|_T^2 \qquad \forall\, T \in \mathcal{T}_{\text{new}} \qquad . \tag{A.2}$$

3. For the next assumption, recall how the predicted stretching direction $\mathbf{p}_i^{\text{pred}}$ and stretching length h_i^{pred} are obtained from $\mathcal{T}_{\text{old}}$. The Hessian strategy and (6.5) in particular imply

   $$(h_i^{\text{pred}}(x))^2 \cdot \left|\partial^{2,h}_{\mathbf{p}_i^{\text{pred}}\mathbf{p}_i^{\text{pred}}}u_h(x)\right| = \alpha_T^2 \qquad i = 1,2,3,\ \forall\, x \in T,\ \forall\, T \in \mathcal{T}_{\text{old}},$$

   i.e. the stretching lengths and the second order derivatives are balanced.

   Then the new mesh $\mathcal{T}_{\text{new}}$ is constructed based on the predicted information. Of course a similar balancing between stretching lengths and second order derivatives should hold there. We replace $\partial^{2,h}u_h$ by ∂^2u and assume that they are similar. The predicted stretching lengths $h_i^{\text{pred}}(x)$ become the actual stretching lengths of the new element $T \in \mathcal{T}_{\text{new}}$. Finally the above relation is not formulated for all x but now in an integral sense and reads

   $$h_{i,T}^2 \cdot \frac{\|\partial^2_{\mathbf{p}_i\mathbf{p}_i}u\|_T}{|T|^{1/2}} \sim \alpha_T^2 \qquad i = 1,2,3, \quad \forall\, T \in \mathcal{T}_{\text{new}}, \tag{A.3'}$$

   with some local element parameters α_T. We stress that this assumption relates the information extracted from the given mesh $\mathcal{T}_{\text{old}}$ (i.e. $\mathbf{p}_i^{\text{pred}}(x), h_i^{\text{pred}}(x)$) to the newly constructed mesh $\mathcal{T}_{\text{new}}$. Only if the new mesh $\mathcal{T}_{\text{new}}$ is close enough to the predicted stretching directions/lengths then (A.3') can be expected to hold.

Numerical experiments have shown that assumption (A.3') can be too strong occasionally. Hence we introduce a weakened version (A.3) which can be derived from (A.3'):

$$h_{min,T}^{-2} \sum_{i=1}^{3} h_{i,T}^4 \cdot \|\partial_{\mathbf{p}_i\mathbf{p}_i}^2 u\|_T^2 \lesssim \sum_{i=1}^{3} h_{i,T}^2 \cdot \|\partial_{\mathbf{p}_i\mathbf{p}_i}^2 u\|_T^2 \quad \forall T \in \mathcal{T}_{\text{new}}. \tag{A.3}$$

For further details see [Kun02b].

4. **The interpolation estimates are sharp.**
   Lemma 6.2.1 states that

   $$\|\partial_{\mathbf{p}_i}(u - \mathrm{I_{CR}}u)\|_T \lesssim \sum_{j=1}^{3} h_{j,T} \cdot \|\partial_{\mathbf{p}_i\mathbf{p}_j}^2 u\|_T \qquad i = 1,2,3.$$

   For a strict proof, we require a certain equivalence of left– and right–hand side (i.e. the converse inequality holds as well). The corresponding assumption becomes

   $$\sum_{i=1}^{3} h_{i,T}^2 \cdot \|\partial_{\mathbf{p}_i\mathbf{p}_i}^2 u\|_T^2 \lesssim \sum_{i=1}^{3} \|\partial_{\mathbf{p}_i}(u - \mathrm{I_{CR}}u)\|_T^2 \quad \forall T \in \mathcal{T}_{\text{new}}. \tag{A.4}$$

5. **$u_h \approx \mathrm{I_{CR}}u$ in the sense of the alignment measure m_1.**
   To transform this into a formula, we simply require

   $$m_1(u - \mathrm{I_{CR}}u, \mathcal{T}_{\text{new}}) \sim m_1(u - u_h, \mathcal{T}_{\text{new}}) \tag{A.5}$$

   because the left–hand side is much easier to investigate. The numerical experiments of [Kun02b] show that this is a realistic demand.

Our main result bounds the alignment measure.

Theorem 6.4.1 (Bounded alignment measure) *Assume that the heuristic assumptions (A.1)–(A.5) hold for all elements of the newly constructed mesh $\mathcal{T}_{\text{new}}$. Then*

$$m_1(u - u_h, \mathcal{T}_{\text{new}}) \sim 1 \qquad . \tag{6.7}$$

Key ideas of the proof: The proof is given in [Kun02b]. Naturally it utilizes the heuristic assumptions which already contain much of the essence of the proof. ∎

Remark 6.4.2 In the *2D case* and with additional heuristic assumptions it is even possible to prove the converse result. If the alignment measure $m_1(u - u_h, \mathcal{T})$ is to be bounded by $\mathcal{O}(1)$ then the maximum stretching ratio and the stretching direction of an element is that of the Hessian strategy.

In other words, the stretching direction and ratio of the Hessian strategy are necessary and sufficient for a bounded alignment measure (2D alone). □

For a more comprehensive discussion of the Hessian strategy and its heuristic assumptions we refer to [Kun02b]. Finally it is worth mentioning that the heuristic assumptions (A.2)–(A.5) also have been analysed *numerically*, see [Kun02b]:

- Assumptions (A.2) and (A.5) did not cause trouble.
- Assumptions (A.3) and (A.4) may be critical, i.e. they are not satisfied for all elements. However, a closer inspection reveals that they are violated only for few elements. Since all assumptions interact in a rather complex way, these violations did not infringe the overall result. This observation could be attributed to the heuristic character of the Hessian strategy.

Bibliography

[AB99]   M. Ainsworth and I. Babuška. Reliable and robust a posteriori error estima-
            tion for singularly perturbed reaction–diffusion problems. *SIAM J. Num. Anal.*,
            36(2):331–353, 1999.

[AD92]   Th. Apel and M. Dobrowolski. Anisotropic interpolation with applications to
            the finite element method. *Computing*, 47:277–293, 1992.

[Ada75]   R. A. Adams. *Sobolev Spaces*. Academic Press, New York, 1975.

[AGJM01]   Th. Apel, S. Grosman, P. K. Jimack, and A. Meyer. A new methodology for
            anisotropic mesh refinement based upon error gradients. Preprint SFB393/01-11,
            TU Chemnitz, 2001.

[AL96]   Th. Apel and G. Lube. Anisotropic mesh refinement in stabilized Galerkin
            methods. *Numer. Math.*, 74(3):261–282, 1996.

[AL98]   Th. Apel and G. Lube. Anisotropic mesh refinement for a singularly perturbed
            reaction diffusion model problem. *Appl. Numer. Math.*, 26:415–433, 1998.

[AN]   Th. Apel and S. Nicaise. On the inf-sup condition for the Bernardi–Fortin–
            Raugel element on anisotropic meshes. Technical report. To appear.

[Ang95]   L. Angermann. Balanced a-posteriori error estimates for finite volume type
            discretizations of convection-dominated elliptic problems. *Computing*, 55(4):305–
            323, 1995.

[ANS01a]   Th. Apel, S. Nicaise, and J. Schöberl. Crouzeix-Raviart type finite elements on
            anisotropic meshes. *Numer. Math.*, 89:193–223, 2001.

[ANS01b]   Th. Apel, S. Nicaise, and J. Schöberl. A non-conforming finite element method
            with anisotropic mesh grading for the Stokes problem in domains with edges.
            IMA J. Numer. Anal., 21:843–856, 2001.

[AO97]   M. Ainsworth and J.T. Oden. A posteriori error estimators for the Stokes and
            Oseen equations. *SIAM J. Num. Anal.*, 34(1):228–245, 1997.

[AO00]   M. Ainsworth and J.T. Oden. *A posteriori error estimation in finite element
            analysis*. Wiley, 2000.

[Ape99]   Th. Apel. *Anisotropic finite elements: Local estimates and applications*. Ad-
            vances in Numerical Mathematics. Teubner, Stuttgart, 1999.

[AR03]   Th. Apel and H. M. Randrianarivony. Stability of discretizations of the Stokes
            problem on anisotropic meshes. *Mathematics and Computers in Simulation*,
            61:437–447, 2003.

[AV02]   A. Agouzal and Yu. Vassilevski. On a discrete Hessian recovery for P_1 finite elements. *J. Numer. Math.*, 10(1):1–12, 2002.

[Bak69]   N. S. Bakhvalov. Optimization of methods for the solution of boundary value problems in the presence of a boundary layer. *Zh. Vychisl. Mat. Mat. Fiz.*, 9:841–859, 1969. In Russian.

[Ban98]   R.E. Bank. A simple analysis of some a posteriori error estimates. *Appl. Numer. Math.*, 26(1-2):153–164, 1998.

[BD97]   G.C. Buscaglia and E.A. Dari. Anisotropic mesh optimization and its application in adaptivity. *Int. J. Numer. Methods Eng.*, 40(22):4119–4136, 1997.

[Bec00]   R. Becker. An optimal control approach to a posteriori error estimation for finite element discretizations of the Navier-Stokes equations. *East-West J. Numer. Math.*, 8(4):257–274, 2000.

[Ber02]   S. Berrone. Robustness in a posteriori error analysis for FEM flow models. *Numer. Math.*, 91(3):389–422, 2002.

[BH96]   F. J. Bossen and P. S. Heckbert. A pliant method for anisotropic mesh generation. In *Proceedings of the 5ᵗʰ Annual International Meshing Roundtable*, Pittsburgh, PA, 1996. Sandia National Laboratories.

[BK94]   R. Beinert and D. Kröner. Finite volume methods with local mesh alignment in 2–D. In *Adaptive Methods – Algorithms, Theory and Applications*, volume 46 of *Notes on Num. Fluid Mechanics*, pages 38–53, Braunschweig, 1994. Vieweg.

[BR78]   I. Babuška and W. C. Rheinboldt. Error estimates for adaptive finite element computations. *SIAM J. Numer. Anal.*, 15(4):736–754, 1978.

[Bur01]   P. Burda. A posteriori error estimates for the Stokes flow in 2D and 3D domains. In P. Neittaanmäki et al., editor, *Finite element methods. Three-dimensional problems*, volume 15 of *GAKUTO Int. Ser., Math. Sci. Appl*, pages 34–44. Gakkotosho, 2001.

[BW85]   R. E. Bank and A. Weiser. Some a posteriori error estimators for elliptic partial differential equations. *Math. Comput.*, 44(170):283–301, 1985.

[BW90]   R. E. Bank and B. D. Welfert. A posteriori error estimates for the Stokes equations: A comparison. *Comput. Methods Appl. Mech. Engrg.*, 82:323–340, 1990.

[CF01]   C. Carstensen and S. Funken. A posteriori error control in low-order finite element discretisations of incompressible stationary flow problems. *Math. Comp.*, 70(236):1353–1381, 2001.

[CHM95]   M.J. Castro-Díaz, F. Hecht, and B. Mohammadi. New progress in anisotropic grid adaption for inviscid and viscous flow simulations. In *Proceedings of the 4ᵗʰ Annual International Meshing Roundtable*, pages 73–85, Albuquerque, NM, 1995. Sandia National Laboratories. Also Report 2671 at INRIA.

[Cia78]   P. G. Ciarlet. *The finite element method for elliptic problems*. North-Holland, Amsterdam, 1978. Reprinted by SIAM, Philadelphia, 2002.

[CKN03]   E. Creusé, G. Kunert, and S. Nicaise.   A posteriori error estima-
          tion for the Stokes problem: Anisotropic and isotropic discretizations.
          Preprint SFB393/03–01, TU Chemnitz, January 2003. Also http://archiv.tu-
          chemnitz.de/pub/2003/0005/index.html.
          Submitted to Math. Models Methods Appl. Sci.

[CR73]   M. Crouzeix and P. A. Raviart. Conforming and non-conforming finite elements
          for solving the stationary Stokes equations. R.A.I.R.O. Anal. Numér., 7:33–76,
          1973.

[DDP95]   E. Dari, R. Durán, and C. Padra.   Error estimators for nonconforming finite
          element approximations of the Stokes problem. Math. Comput., 64(211):1017–
          1033, 1995.

[DGP99]   M. Dobrowolski, S. Gräf, and C. Pflaum.   On a posteriori error estimators
          in the finite element method on anisotropic meshes. Electronic Transactions
          Num. Anal., 8:36–45, 1999.

[Dol98]   V. Dolejší. Anisotropic mesh adaptation for finite volume and finite element
          methods on triangular meshes. Comput. Vis. Sci., 1(3):165–178, 1998.

[FG00]   P.J. Frey and P.L. George. Mesh generation. Application to finite elements.
          Hermes Science Publishing, Oxford, 2000.

[Fle99]   P. Fleischmann. Mesh Generation for Technology CAD in Three Dimensions.
          PhD thesis, Technische Universität Wien, 1999.

[FLR96]   J. Fröhlich, J. Lang, and R. Roitzsch. Selfadaptive finite element computations
          with smooth time controller and anisotropic refinement. Report 96–16, ZIB,
          1996.

[FP00]   L. Formaggia and S. Perotto. Anisotropic error estimates for elliptic problems.
          Report 18, Ecole Polytechnique Federale de Lausanne, 2000.

[FP01]   L. Formaggia and S. Perotto. New anisotropic a priori error estimates. Numer.
          Math., 89(4):641–667, 2001.

[FPZ01]   L. Formaggia, S. Perotto, and P. Zunino. An anisotropic a-posteriori error esti-
          mate for a convection-diffusion problem. Comput. Vis. Sci., 4(2):99–104, 2001.

[GR86]   V. Girault and P.-A. Raviart. Finite element methods for Navier-Stokes equa-
          tions, Theory and algorithms, volume 5 of Springer Series in Computational
          Mathematics. Springer, Berlin, 1986.

[Gro02]   S. Grosman. Robust local problem error estimation for a singularly perturbed
          reaction-diffusion problem on anisotropic finite element meshes. Technical report
          SFB393/02–07, TU Chemnitz, May 2002.

[HDB⁺00] W.G. Habashi, J. Dompierre, Y. Bourgault, D. Ait-Ali-Yahia, M. Fortin, and
          M.-G. Vallet. Anisotropic mesh adaptation: Towards user-independent, mesh-
          independent and solver-independent CFD solutions: Part I: General principles.
          Int. J. Numer. Methods Fluids, 32:725–744, 2000.

[HL98]   R. Hangleiter and G. Lube. Stabilized Galerkin methods and layer-adapted grids
          for elliptic problems. Comput. Methods Appl. Mech. Eng., 166(1–2):165–182,
          1998.

[JL00]   J. Jou and J.-L. Liu. An a posteriori finite element error analysis for the Stokes
            equations. *J. Comput. Appl. Math.*, 14(2):333–343, 2000.

[Joh98]   V. John. A posteriori L^2-error estimates for the nonconforming P_1/P_0-finite ele-
            ment discretization of the Stokes equations. *J. Comput. Appl. Math.*, 96(2):99–
            116, 1998.

[Joh00]   V. John. A numerical study of a posteriori error estimators for convection–
            diffusion equations. *Comput. Methods Appl. Mech. Engrg.*, 190:757–781, 2000.

[KLR02]   T. Knopp, G. Lube, and G. Rapin. Stabilized finite element methods with shock
            capturing for advection-diffusion problems. *Comput. Methods Appl. Mech. Eng.*,
            191(27–28):2997–3013, 2002.

[Kop01]   N. Kopteva. Maximum norm a posteriori error estimates for a one-dimensional
            convection-diffusion problem. *SIAM J. Numer. Anal.*, 39(2):423–441, 2001.

[KR90]   R. Kornhuber and R. Roitzsch. On adaptive grid refinement in the presence
            of internal and boundary layers. *IMPACT of Computing in Sci. and Engrg.*,
            2:40–72, 1990.

[Kra01]   J. Krause. *On boundary conforming anisotropic Delaunay meshes.* PhD thesis,
            ETH Zürich, 2001. Also published at Hartung-Gorre Verlag.
            http://e-collection.ethbib.ethz.ch/show?type=diss&nr=14219.

[KS00]   D. Kay and D. Silvester. A posteriori error estimation for stabilized mixed
            approximations of the Stokes equations. *SIAM J. Sci. Comput.*, 21(4):1321–
            1336, 2000.

[KS01]   D. Kay and D. Silvester. The reliability of local error estimators for convection–
            diffusion equations. *IMA J. Numer. Anal.*, 21(1):107–122, 2001.

[Kun99]   G. Kunert. *A posteriori error estimation for anisotropic tetrahedral and trian-
            gular finite element meshes.* Logos Verlag, Berlin, 1999. Also PhD thesis, TU
            Chemnitz, http://archiv.tu-chemnitz.de/pub/1999/0012/index.html.

[Kun00]   G. Kunert. An a posteriori residual error estimator for the finite element method
            on anisotropic tetrahedral meshes. *Numer. Math.*, 86(3):471–490, 2000. DOI
            10.1007/s002110000170.

[Kun01a]   G. Kunert. A local problem error estimator for anisotropic tetrahedral finite
            element meshes. *SIAM J. Numer. Anal.*, 39(2):668–689, 2001.

[Kun01b]   G. Kunert. A note on the energy norm for a singularly perturbed model problem.
            Preprint SFB393/01–02, TU Chemnitz, January 2001. Also http://archiv.tu-
            chemnitz.de/pub/2001/0006/index.html.

[Kun01c]   G. Kunert. A posteriori H^1 error estimation for a singularly perturbed reaction
            diffusion problem on anisotropic meshes. *Submitted to IMA J. Numer. Anal.*,
            2001.

[Kun01d]   G. Kunert. A posteriori H^1 error estimation for a singularly perturbed reaction
            diffusion problem on anisotropic meshes. Preprint SFB393/01–21, TU Chemnitz,
            August 2001.
            Also http://archiv.tu-chemnitz.de/pub/2001/0073/index.html.

[Kun01e]   G. Kunert. Robust a posteriori error estimation for a singularly perturbed reaction–diffusion equation on anisotropic tetrahedral meshes. *Adv. Comp. Math.*, 15(1–4):237–259, 2001.

[Kun01f]   G. Kunert. Robust local problem error estimation for a singularly perturbed problem on anisotropic finite element meshes. *Math. Model. Numer. Anal.*, 35(6):1079–1109, 2001.

[Kun02a]   G. Kunert. A note on the energy norm for a singularly perturbed model problem. *Computing*, 69(3):265–272, 2002.

[Kun02b]   G. Kunert. Towards anisotropic mesh construction and error estimation in the finite element method. *Numer. Meth. PDE*, 18(6):625–648, 2002.

[Kun03]   G. Kunert. A posteriori error estimation for convection dominated problems on anisotropic meshes. *Math. Methods Appl. Sci.*, 26(7):589–617, 2003.

[KV00]   G. Kunert and R. Verfürth. Edge residuals dominate a posteriori error estimates for linear finite element methods on anisotropic triangular and tetrahedral meshes. *Numer. Math.*, 86(2):283–303, 2000. DOI 10.1007/s002110000152.

[Lin00]   T. Linß. Uniform superconvergence of a Galerkin finite element method on Shishkin-type meshes. *Numer. Meth. PDE*, 16(5):426–440, 2000.

[Lin03]   T. Linß. Layer–adapted meshes for convection–diffusion problems. *Comput. Methods Appl. Mech. Eng.*, 192(9–10):1061–1105, 2003.

[LS01]   T. Linß and M. Stynes. The SDFEM on Shishkin meshes for linear convection-diffusion problems. *Numer. Math.*, 87(3):457–484, 2001.

[Mel02]   J.M. Melenk. *hp-Finite Element Methods for Singular Perturbations*. Lecture Notes in Mathematics. Springer, Heidelberg, 2002. Habilitation thesis.

[Mor96]   K. W. Morton. *Numerical solution of convection-diffusion problems*. Chapman & Hall, London, 1996.

[MOS96]   J.J.H. Miller, E. O'Riordan, and G.I. Shishkin. *Fitted numerical methods for singularly perturbed problems. Error Estimates in the maximum norm for linear problems in one and two dimensions*. World Scientific Publications, Singapore, 1996.

[Noc95]   R.H. Nochetto. Pointwise a posteriori error estimators for elliptic problems on highly graded meshes. *Math. Comp.*, 64:1–22, 1995.

[Pic03]   M. Picasso. An anisotropic error indicator based on Zienkiewicz-Zhu error estimator: Application to elliptic and parabolic problems. *SIAM J. Sci. Comput.*, 24(4):1328–1355, 2003.

[PV00]   A. Papastavrou and R. Verfürth. A posteriori error estimators for stationary convection–diffusion problems: A computational comparison. *Comput. Methods Appl. Mech. Eng.*, 189(2):449–462, 2000.

[PVMZ87]   J. Peraire, M. Vahdati, K. Morgan, and O. C. Zienkiewicz. Adaptive remeshing for compressible flow computation. *J. Comp. Phys.*, 72:449–466, 1987.

[Rac93]   W. Rachowicz. An anisotropic h-type mesh refinement strategy. *Comput. Methods Appl. Mech. Engrg.*, 109:169–181, 1993.

[Ran01]   M. Randrianarivony. Strengthened Cauchy inequality in anisotropic meshes and application to an a-posteriori error estimator for the Stokes problem. Preprint SFB393/01–23, TU Chemnitz, September 2001.

[RGK93]   W. Rick, H. Greza, and W. Koschel. FCT-solution on adapted unstructured meshes for compressible high speed flow computations. In E.H. Hirschel, editor, *Flow simulation with high-performance computers I*, volume 38 of *Notes on Num. Fluid Mechanics*, pages 334–438. Vieweg, 1993.

[RL99]   H.-G. Roos and T. Linß. Sufficient conditions for uniform convergence on layer-adapted grids. *Computing*, 63(1):27–45, 1999.

[RL01]   H.-G. Roos and T. Linß. Gradient recovery for singularly perturbed boundary value problems II: Two-dimensional convection-diffusion. *Math. Models Meth. Appl. Sci.*, 11(7):1169–1179, 2001.

[Roo98]   H.-G. Roos. Layer-adapted grids for singularly perturbed boundary value problems. *Z. Angew. Math. Mech.*, 78(5):291–309, 1998.

[RST96]   H.-G. Roos, M. Stynes, and L. Tobiska. *Numerical methods for singularly perturbed differential equations. Convection-diffusion and flow problems.* Springer, Berlin, 1996.

[San01]   G. Sangalli. A robust a posteriori estimator for the Residual–free Bubbles method applied to advection-diffusion problems. *Numer. Math.*, 89(2):379–399, 2001.

[SE00]   Y.-T. Shih and H.C. Elman. Iterative methods for stabilized discrete convection diffusion problems. *IMA J. Numer. Anal.*, 89(3):333–358, 2000.

[Sie96]   K.G. Siebert. An a posteriori error estimator for anisotropic refinement. *Numer. Math.*, 73(3):373–398, 1996.

[Sim94]   R.B. Simpson. Anisotropic mesh transformation and optimal error control. *Appl. Numer. Math.*, 14:183–198, 1994.

[SR99]   T. Skalický and H.-G. Roos. Anisotropic mesh refinement for problems with internal and boundary layers. *Int. J. Numer. Meth. Engrg.*, 46:1933–1953, 1999.

[SSS99]   D. Schötzau, Ch. Schwab, and R. Stenberg. Mixed *hp*-FEM on anisotropic meshes II: Hanging nodes and tensor products of boundary layer meshes. *Numer. Math.*, 83:667–697, 1999.

[ST97]   M. Stynes and L. Tobiska. Error Estimates and Numerical Experiments for Streamline-Diffusion-Type Methods on Arbitrary and Shishkin Meshes. *CWI Quarterly*, 10(3–4):337–352, 1997.

[ST01]   M. Stynes and L. Tobiska. Analysis of the streamline-diffusion finite element method on a piecewise uniform mesh for a convection-diffusion problem with exponential layers. *East-West J. Numer. Math.*, 9(1):59–76, 2001.

[Ver89]   R. Verfürth. A posteriori error estimators for the Stokes equation. *Numer. Math.*, 55:309–325, 1989.

[Ver91]   R. Verfürth.   A posteriori error estimators for the Stokes equation II: Non-conforming discretizations. *Numer. Math.*, 60:235–249, 1991.

[Ver94]   R. Verfürth.   A posteriori error estimation and adaptive mesh–refinement techniques. *Journal of Computational and Applied Mathematics*, 50:67–83, 1994.

[Ver96]   R. Verfürth.   *A review of a posteriori error estimation and adaptive mesh–refinement techniques*. Wiley-Teubner, Chichester; Stuttgart, 1996.

[Ver98a]   R. Verfürth.   A posteriori error estimators for convection–diffusion equations. *Numer. Math.*, 80(4):641–663, 1998.

[Ver98b]   R. Verfürth.   Robust a posteriori error estimators for singularly perturbed reaction–diffusion equations. *Numer. Math.*, 78(3):479–493, 1998.

[VH96]   R. Vilsmeier and D. Hänel.   Computational aspects of flow simulation in three dimensional, unstructured, adaptive grids. In E.H. Hirschel, editor, *Flow simulation with high-performance computers II*, volume 52 of *Notes on Num. Fluid Mechanics*, pages 431–446. Vieweg, 1996.

[XF03]   Chr. Xenophontos and S.R. Fulton.   Uniform approximation of singularly perturbed reaction-diffusion problems by the finite element method on a Shishkin mesh. *Numer. Meth. PDE*, 19(1):89–111, 2003.

[ZR93]   G. Zhou and R. Rannacher.   Mesh orientation and anisotropic refinement in the streamline diffusion method. In M. Křížek, P. Neitaanmäki, and R. Stenberg, editors, *Finite Element Methods: Fifty Years of the Courant Element*, volume 164 of *Lecture Notes in Pure and Applied Mathematics*, pages 491–500. Marcel Dekker, Inc., New York, 1993. Also published as Preprint 93-57, Universität Heidelberg, IWR, SFB 359, 1993.

[ZW94]   O. C. Zienkiewicz and J. Wu.   Automatic directional refinement in adaptive analysis of compressible flows. *Int. J. Numer. Methods Engrg.*, 37:2189–2210, 1994.

Index